GROWING A
HOUSEBUILDER

Nourishing Spirit
Through an Age-Old Art

Mark Brady

Lifetime Learning Series

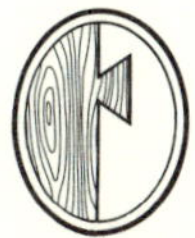

Dovetail Publishers, Inc.
Menlo Park, California

Library of Congress Cataloging in Publication Data
Brady, Mark
Growing a Housebuilder
Bibliography
Index
1. Housebuilding 2. Self-development 3. Education
4. Construction 5. Architecture 6. Work

Library of Congress LC89-040567

ISBN 0-9624345-3-1

First Printing February, 1990
93 92 91 90 11 10 9 8 7 6 5 4 3 2 1

Printed in the United States of America

Table of Contents

"As if the story of a house were told, or ever could be."
— E. A. Robinson

For Susan

Many people contribute to a work of this nature. Pablo Saravia took on my building duties spectacularly. Duane Elgin took on editing duties superlatively. Others I wish to acknowledge here are Robert Lawrence Riley who gave me my first work in housebuilding. May he rest in gentle peace. Eric LeVesque gave me my second work. Dave Gilbreath provided the support and encouragement necessary to undertake my first work in business. Robert Rubenstein did the same for my second. And to all those other special people who have encouraged me in the many other meaningful first and second experiences of my life, to all of them I wish to extend my sincerest thanks and appreciation.

Preface

This book chronicles the planning and building of a house for sale on speculation ("New House") in the City of Palo Alto, California in 1986-87.[1] When a person undertakes to build something in the real world two conditions are required: action and information. The majority of the books that address the question "What's required to build a house?" provide mostly technical information. These books are reminiscent of the Sufi Nasrudin, who, when asked what his house was like, brought the questioner a single brick and replied, "My house is just a collection of these."[2] Such books only address the lumber mechanics of building. They omit a lot of the real action, the physical, emotional, mental, spiritual and community aspects intricately interwoven into the construction of any building. This book delves into these experiences and demonstrates how they all contribute to "growing a housebuilder".

While I have more than an abiding interest in things psychological and spiritual, except for how a decade of formal and informal study has molded and shaped my character, few spiritual or psychological techniques were ever intentionally introduced to the people who worked on New House. At no point did I gather the plumbers, sheetrockers, carpenters and bankers to hold hands together in a circle, close their eyes and chant "I love Freud" or "God is my Savior". However, I did pay close attention to my own internal and external experience on a daily basis during the course of this building project. It is that experience and the accompanying insights that I present herein.

Growing a Housebuilder is not written only for people who do construction for a living. It will be of interest to all people who experience their work as an arena for growth and development — practical-minded, results-oriented people open to discovering the learning that this work in particular, and all work in general, have in common. It is intended to be a work of service to people building their own unique approach to life and living.

My intention is that *Growing a Housebuilder* serve by acknowledging that taking risks in the work we choose to do is often frightening and difficult. Making the transition from carpenter to housebuilder, or from general contractor to developer, while they might be viewed as growing, evolutionary steps, they also have a dark and difficult side to them. This book is intended to be a resource for people in any field consciously wanting to make such moves. Even though the work might be difficult and fear-filled, the challenges can be met; even though we may feel alone and unsupportable, there are people and resources that will support us. (Many are identified throughout.) For those of us without family, friends can substitute. For those unable to rely upon friends, therapists and therapeutic materials can be valuable supportive allies. For contractors, allies can be other tradesmen or building officials. I hope I have demonstrated that help is indeed available when I am willing and wish to receive it. I also contend that it is possible to do this work in a much larger context, as a daily learning and teaching process, to be working in the world while remaining apart from it (at least some of the time).

In my life I have frequently been impressed with the fact that musicians are able to use music, rhyme and rhythm to put forth personal and spiritual truths; that media artists are able to use paint and canvas to attempt the same. Only they know how near or far from their intended vision a completed work approaches. As such, the following pages might be considered a journeyman carpenter, using tangible and intangible resources to investigate and express similar personal truths. I leave you, the reader, the opportunity to assess my attempts for yourself.

Mark Brady
Atherton, California

November 23, 1989

1

"As we respond with caring and vision to
all work, we develop our capacity to
respond fully to all of life."
— Tarthang Tulku

First Works

I never originally intended to make a career in construction. In order to take my first construction job I had to resign the vice-presidency and co-ownership of a profitable and well-connected manufacturing company I had founded with a friend only four years before. I also had to give up weekly paychecks that were so large that I only cashed one a month. Also abandoned was a company car, boat and airplane and a company credit card for stores and restaurants all over town. Simultaneously, I withdrew as a full-time student at UCLA. My company's products were war related, and my UCLA studies were mostly antiwar, and I was mostly becoming schizophrenic. So I abruptly decided to chuck it all in exchange for getting up and punching-in at 7 AM to begin work at four dollars an hour. It wasn't a rational decision, but it was a right one.

It was also an ignorant decision, for I had absolutely no idea how much physical pain the body is capable of experiencing while keeping a person alive and conscious the whole time. After many years of four and five boring-hour workdays behind a desk, my first job in construction was "packing" lumber. I picked up and carried sixteen and twenty foot long 2 x 10's and 2 x 12's and placed them up onto the second story of a furniture-store-in-the-making. Eight hours of packing lumber generates horrendous pain even in well-conditioned construction workers. In me it generated twisted, living death. Somehow I managed to make it through my first day, only to show up on the job the next to do the same thing all over. On the third day the Gods showed mercy. It rained. I stayed home, soaked my bruised and aching body, and read framing manuals the whole day. In spite of the pain and having to work a real eight hour workday, something in me really responded to the job and to the people.

I helped complete the furniture store project in Southern California and worked on several other jobs after that. Less than six months later I was being sent out with my own crews to do small jobs like garages, carports, fences and bedroom additions. And I loved it.

But I didn't love Southern California. So it wasn't long before I moved back to the east coast from where I had been living ten short years before. I kicked around awhile, not sure about what direction I should take. Back to school? Back to building? Back to manufacturing? Then one day, seemingly by chance, I discovered that Derek*, a friend from high school, had just purchased a building lot in a small town in upstate New York. He and another friend had decided to go into the house-building business, in the town of New Paltz, as a new company called Bountiful Builders. But since neither of them had very much experience, I was a welcome $6 an hour, six-months-experienced addition to the crew. I also had a 50% raise.

Not surprisingly, we didn't build the greatest first house. But we did get it built and it did get sold. It was built on a foundation prepared and poured by professionals so it's still standing today, albeit with some cabinets a bit askew, some floors fit for downhill skiing, and the roof shingle courses tending to ride off into the sunset. Later I learned that it is just these personal touches that lend a custom home its unique color and character!

Even with our errors and inexperience, the work was inspiring. I never felt so alive. I would be on the job by 6 AM, work through to 5 with only a lunch break, have dinner and then work without pay on Derek's personal house until it got so dark that I had to guess where the nails were to hit them. It was great to be outdoors, surrounded by the seasons as they went through their changes, while simultaneously the houses under construction went through theirs, and I went through mine. Even winter-building had its own feel (cold) and flavor (cold). Still, I felt so energized

* *All of the characters in this book are real people or composites drawn from different people I have known. Names have been changed except where noted.*

and excited during this time of learning, I'd be out in the morning licking ice from my beard and shoveling snow off the deck hours before the sun came up. This was an intense, magical time in my life — a time for learning and growing, a time for great risk-taking. It was a young time when all of life felt plumb, level and proper.

One change saw me resuming college studies. At UCLA I had majored in English. In New Paltz I had done construction work for several psychology professors. Because I enjoyed working with them, I found myself drawn to their classes. I switched majors, took the necessary courses and graduated in two years with several semesters of honors.

I loved learning about my own inner workings, only now I found myself with two works that felt right for me, housebuilding and psychology. By this time I had helped build more than a dozen dwellings from scratch, and I felt like there was little new to learn about the mechanics of housebuilding. But there seemed to be a whole world of learning left concerning the mechanics of me-building. So I took a sabbatical from homebuilding and went off in search of "higher self-knowledge". Ironically, it was during that sabbatical I learned a huge amount about housebuilding. Not the lumber-mechanics of construction, but the "me-mechanics" of the work. By the me-mechanics I mean the psychological, social, mental and emotional aspects that arise naturally during the complex process of constructing a new house.

There were many advantages in learning the lumber-mechanics of housebuilding as an employee for Bountiful

Builders. First and foremost, they paid for my introduction to the complete housebuilding process. In return, they received the benefit of my energy, enthusiasm and the inevitable mistakes that a novice builder makes. On and off the job our relationships were cooperative and collegial and the work was great fun. In retrospect the fact that I preferred to be an employee rather than a partner or boss told a great deal about my initial attitude about building. As an employee I was willing to be responsible for coming to work on time, for working hard and well, and for directing helpers with the work at hand. In short, I was willing to learn and to grow and work to consistently build the best house that I could. I was not willing to hire and fire, deal with owners and buyers, or be responsible for the financial pressures I knew accompany any fledgling company. I suspected that one day I might be willing to accept those responsibilities, but not as a novice builder working with an undercapitalized and inexperienced company like Bountiful Builders.

I learned much from Derek's and his partner's mistakes and learnings. Several episodes stand out clearly in mind.

After Bountiful Builders completed and sold its first six houses and the business began to earn a profit, Derek felt an inner drive to begin a college education that he, like me, had postponed for half a decade. With my suggestions and experience as a guide, he enrolled in several of the psychology courses that I found to be most provocative and challenging. Unlike me, however, he seemed to breeze through the classes with minimal effort, attaining straight A's his first semester. Confident that he could do the work required of an undergraduate, Derek decided to work towards becoming a doctor. Medicine would be his chosen profession — his way to be of service in the world. This seemed both admirable and entirely within the realm of possibility.

It so happened, too, that Derek was a spiritual seeker — a member of a small group who gathered each week to study Sufism, a school of esoteric spiritual teachings derived from Islamic mysticism. Derek was very dedicated to this group and to these teachings (which required saying prayers twice daily and abstaining from food during the day for one week each year). He was also very devoted to the 70 year old Iranian originator of this school who visited the group every two or three years. Shortly after Derek had enrolled in his second college semester at New Paltz, his Sufi teacher came to the United States for a visit and Derek related his plans to become a doctor.

"Do not become a doctor," his teacher advised. "Instead, continue to provide shelter for people."

The next day Derek withdrew from college and totally rededicated his life to housebuilding. At the time of Derek's transition, I found it very difficult to understand. He seemed excited and resolved in his decision to become a doctor, and yet he was able to abandon it overnight and return to housebuilding — work that I imagined he felt full and finished with. And all this on the simple directive from someone he encountered only three or four times per decade!

My fantasy is that Derek's Sufi teacher was familiar with this story:

In medieval times if you wanted to begin a spiritual

path and were lucky enough to meet somebody who might be a teacher (you never knew for sure), that person would say to you, "What is it you wish to do for a living?" The seeker might say, "Well, I want to be a housebuilder." The teacher would tell him, "There's a wonderful housebuilder in New Paltz who may be somebody quite spiritual. I'll send you to him and you can apprentice with him. The only thing you must remember is to never, ever discuss anything spiritual with him."

And so off the seeker would go and work for the housebuilder for five or six years as an apprentice, and he would really get to know how to build houses. Finally the housebuilder would tell the seeker, "You know, you really know how to build houses now. There is an opportunity in Atherton to become partners with a very special man who builds houses there. I can get you into that partnership, but you must never, ever discuss anything spiritual with that man. You must concentrate on making very, very beautiful houses." And so off the seeker would go to Atherton and work for many years, and shortly before his death, the aged partner would tell the seeker, "Soon this business will be yours. You've become a really fine housebuilder. One of these days someone who is younger may come wanting to learn something spiritual from you. Only teach him about building houses."[1]

I was later to learn that the teacher's reasons were much less mysterious: shortly before making his pronouncement to Derek, he had undergone a serious operation which the doctors had severely botched. Fresh from an unnecessarily prolonged recovery, the teacher delivered that advice. In spite of this, Derek persisted and learned to build very fine houses. He has since gone on to develop a distinguished list of clients that includes the Rockefellers, chief executives of major U. S. Corporations, and heads of State.

I learned important lessons in the economics of building when, in the Spring of '75, home construction in New Paltz took a downturn. There probably aren't more than 15 or 20 new homes built there in the best upturned year, but that spring Bountiful Builders built none. To make matters worse, one client decided he didn't like the home we'd built him and filed suit for a number of specious reasons (he didn't like the plywood siding he had authorized us to install and there were scuff marks on his bathroom and kitchen floor!). Because Bountiful Builders operated primarily on a handshake and good intentions, little beyond the original construction contract was ever committed to paper. With so little documentation, the client managed to win a substantial judgement in court. This proved to be money Bountiful Builders didn't have, and so Derek and his partner filed for bankruptcy. The situation went from bad to worse when I learned about the additional forty thousand dollars they didn't have that was owed to subcontractors and suppliers. Materials and labor that I had happily provide by request, were unhappily included in that shortfall.

Since my own personal life was seriously undercapitalized at the time, this money was of some conse-quence. Needless to say I never did see all of it. Derek's partner swiftly (and surreptitiously) found a buyer for the house we'd built for him and promptly found a great job offer in another part of the state.

Like any good builder, Derek however decided to hang in there. While not able to assume complete responsibility for all of the company's debts alone, he did take responsi-bility for half of them — fifty percent of every dollar owed was paid out of his personal assets to all subcontractors and suppliers. As most participants in bankruptcies know well, fifty cents on the dollar is a real bonanza.

As a very interested participant-observer, I learned a number of things from this experience. The first is that new homeowners can get over the romance stage of the re-lationship very quickly. So shake hands, be nice, but get all contract changes in writing. The other thing I learned is that loans returning only fifty cents on the dollar violate an important rule of fiscal responsibility: preserve capital. Even when those loans are made to friends with high spiritual aspirations, it doesn't hurt to get a personal note and collateral.

Shortly after Bountiful's bankruptcy ending, Derek's and my plans took us both away from New Paltz. He left for New York City for bigger and better building, and I headed for the San Francisco Bay Area to continue my own growth as a housebuilder.

2

"It's what God put a tongue in your mouth
for. If it don't taste right, spit it out."
— Nick Lindsay, *Carpenter*

Preplanning Preparations

Once settled on the San Francisco peninsula, necessity raises a desire to build my own first house on speculation. Until this time I have been able to keep busy with roofing and remodeling work, and I have every reason to believe that more work will continue to come along as it always has in the past. But when my current work runs out, no other work comes. I start advertising and calling friends. Still no work comes and the bills are beginning to pile up. Since all the other contractors I know have plenty of work, it's clear that something is not right and needs changing. Being the psychologically oriented person I am now trained to be, I assume that the something needing to change is me.

My assumption is that a part of me does not want to work. When I ask myself what some of the reasons might be, I come up with a large batch all at once: I don't like taking orders. I don't like being responsible to people I may or may not like. I don't like regular hours. The work is hard. I find it difficult to meet new people. It's especially stressful for me to ask people for money. It's demanding to plan and focus all the various tasks involved in construction. Finally, the work is physically hard and I resent the large pains and small hurts that come with each day's labor.

This is more revelation than I bargained for and I hesitate to inquire further. Still I persist and investigate why I have been only doing small jobs like reroofs, kitchen and bath remodels and additions, using no more than one or two employees. What I discover is not easy to accept: I don't feel emotionally old enough to do larger jobs. I'm not tough enough. My span of attention is too short. I can't hack the financial pressure. I lack technical knowhow. I dislike directing and supervising other people. I have little ability to correct peoples' mistakes with clarity and peace. I don't want the responsibility of having to keep other people employed full-time when I can't even do a good job of keeping myself working. Because I have to do it for a living, construction is not fun any more.

After allowing all my fears their voice, in an attempt to balance the investigation, I decide to investigate my building strengths: I'm good at it. I do nice work. I have a good reputation. I have a lot of experience. I can solve problems quickly and creatively. I can spot problems early and analyze ramifications all up and down the line. I'm fast and efficient. I'm well-organized and good at planning smaller jobs. I like doing work outdoors with a physical exercise factor already built right in.

Although this analysis is clarifying, I still have no work, no prospects and no money. Part of me still wonders when "things" are going to change. The larger part wonders when *I* am going to change. My desire is rising.

"When the time is right," my wife, Susan, says to me one night. "When the need becomes sufficient. When the time is right and the need sufficient, the appropriate actions are taken." Her comments remind me of a story Derek once told me:

The Increasing of Necessity

The tyrannical ruler of Turkestan was listening to the tales of a teacher one evening, when he bethought himself of asking about Khidr.

"Khidr", said the teacher, "comes in response to need. Seize his coat when he appears and all-knowledge is yours."

"Can this happen to anyone?" asked the king.

"Anyone capable," said the teacher.

"Who more 'capable' than I?" thought the king, and he published a proclamation:

"He who presents to me the Invisible Khidr, the Great Protector of Men, him shall I enrich."

A poor old man by the name of Bakhtiar Baba, hearing this cried by the heralds, formed an idea. He said to his wife:

"I have a plan. We shall soon be rich, but a little

later I shall have to die. But this does not matter, for our riches will leave you well provided for."

Then Bakhtiar went before the king and told him that he would find Khidr within forty days, if the king would give him a thousand pieces of gold. "If you find Khidr," said the king, "you shall have ten times this thousand pieces of gold. If you do not, you will die, executed at this very spot as a warning to those who trifle with kings."

Bakhtiar accepted the conditions. He returned home and gave the money to his wife, as a provision for the rest of her life. The rest of the forty days he spent in contemplation, preparing himself for the other life.

On the fortieth day he went before the king. "Your Majesty," he said, "your greed caused you to think the money would produce Khidr. But Khidr, as it is related, does not appear in response to something given from a position of greed."

The king was furious. "Wretch, you have forfeited your life: who are you to trifle with the aspirations of a king?"

Bakhtiar said: "Legend has it that any man may meet Khidr, but the meeting will be fruitful only in so far as that man's intentions are correct. Khidr, they say, would visit you to the extent and for the period that you were worth his while being visited. This is something over which neither you nor I has any control."

"Enough of this wrangling," said the king, "for it will not prolong your life. It only remains to ask the ministers assembled here for their advice upon the best way to put you to death."

He turned to the First Minister and said: "How shall this man die?"

The First Minister said: "Roast him alive as a warning."

The Second Minister, speaking in order of precedence, said: "Dismember him limb from limb."

The Third Minister said: "Provide him with the necessities of life, instead of forcing him to cheat in order to provide for his family."

While this discussion was going on, an ancient sage had walked into the assembly hall. As soon as the Third Minister had spoken, the sage said: "Every man thinks and speaks in accordance with his permanent hidden prejudices."

"What do you mean?" asked the king.

"I mean that the First Minister was originally a baker, so he speaks in terms of roasting. The Second Minister used to be a butcher, so he talks about dismemberment. The Third Minister, having made a study of statecraft, sees the true origin of the matter we are discussing.

"Note two things. First, that Khidr appears and serves each man in accordance with that man's ability to profit by his coming. Second, that this man Bakhtiar, whom I name Baba in token of his sacrifices, was driven by despair to do what he did. He increased his necessity and accordingly he made me appear to you."

As they watched, the ancient sage melted before

their eyes. Trying to do what Khidr directed, the king gave a permanent allowance to Bakhtiar. The first two Ministers were dismissed, and the thousand pieces of gold were returned to the royal treasury by the grateful Bakhtiar Baba and his wife.

How the king was able to see Khidr again, and what transpired between them is in the story of the story of the story of the Unseen World.[1]

Recalling this story, it is clear that I am not ready to "increase my necessity" to the point of death so that my wife can obtain the sufficient insurance riches. But once the calls start coming in heavily from people to whom I owe money, I find my necessity sufficiently increased to begin thinking seriously about taking on bigger jobs.

About the time I finally do find my necessity sufficiently increased, the no-money-down house-buying wave is cresting across America. On the San Francisco Peninsula there are books in every bookstore window, hour long TV "infomercials" on all the obscure local channels (Cable TV hasn't made it to this part of the world yet), and "free" $295 seminars in all the major hotel conference rooms.

With not much to do one Sunday morning around 5:30 AM, I happen to flip over to Mr. Ted Barkley, a real estate tycoon, nestled in amongst the other TV Evangelists singing the praises of the Lord who has made them rich. Ted's message is a little different: he's going to show me how to buy property with no money down as a way of serving God and my fellow man. And after that he's going to teach me to live by the Holy Principles of the Three R's: Rolex, Rolls Royce and Retirement. From the state of my impoverished mentality, Ted's message has great appeal. Ted's a real upbeat guy, very excited about how easy it all is. Buy property with no money down. It sure makes sense to hear him tell it. It makes even more sense that *I* should be doing it, since even Dwayne (one of Ted's very first students who dropped out of High School in the third grade) has bought property on the cheap over and over again.

Out of Ted's hour-long presentation, one fact stands out with particular clarity: Dwayne knows how to perservere. To be in school for fifteen years and only make it through the third grade requires enormous patience and fortitude. The question is, "Do I have what Dwayne has?" Do I also have the $295 for the seminar. The answer is "no" to both. But I do have $17.95 for Robert Allen's book, *Nothing Down*. And I also have my wife for more friendly, helpful advice. Our conversations go something like this:

"How can I buy houses for no money down when anything decent under $250,000 on the Peninsula gets two and three back up offers before it even hits the Multiple Listing Service?"

"So don't buy for no money down. Make a down payment."

"Make a down payment? What a radical idea. Unfortunately I can't even make my truck payment this month."

"Then borrow the down payment. Many people are willing to invest in assets with good upside risk."

"Now that you mention it, I do know a lot of people with a lot of money. Why not invest in me? I know the market. I know how to build. I'm honest and I can get the job done."

This conversation serves to remind me of a well-known book on the laws of money. It's the first law actually: The First Law of Money: Do it! Money Will Come When You Are Doing The Right Thing.[2]

However money is only one of many worries. Before I can "do it", before I can really build a house on spec, I have to make some changes first. In his book on personal development, *Be All That You Are*, Jim Fadiman tells me directly that I can become no more than I have imagined myself capable of being, and that my opinion of myself is accurately reflected in the level of my accomplishments. (Groan.) He then details four essential areas where I can begin to make changes: motivation, awareness, knowledge and practice.[3] The first two I possess. I know what I want and I know that I want it: I want to build a house for sale on speculation. Further, I have the building knowledge I need, along with the recognition that in building it is critical not to dupe myself into thinking I know something that I don't. As I already know, but shall later discover once again, such lapses in awareness can be disturbing and very expensive.

The knowledge I don't have is of more concern. I don't know all the legal aspects and ramifications of land-buying. I don't know all the Uniform Building Codes, or the local city and county ordinances. I don't know the best subcontractors or material suppliers to use. I don't know where to get all the money I'll need to build New House.

Of even greater concern to me are those things that I don't know I don't know. It is this concern which generates the anxiety that Joseph Pearce has called "the enemy of all intelligence".[4] Ignorance of the unknown and a fear that circumstances can arise that I won't be able to handle — these serve as the largest stumbling blocks on the road to becoming a first-time spec-builder.

But necessity is a powerful mother. I am determined to move forward. The first step is to acquire some knowledge in those areas where I know there is a lack. I talk to other contractors about material suppliers and subs. I talk to lawyers and realtors about real estate, get copies of local ordinances, and read books about financing. I even buy an abridged copy of the Uniform Building Code (well, abridged *is* a beginning). The learning feels good. It confirms my desire to move forward.

So much for outer resources. I begin to work simultaneously on inner resources. One of the first journeyman I ever worked with as an apprentice summarized my situation very simply with the statement: "You are what you think about all day long." At this point in my life, my daily thinking is very negative. When I realize the extent and subtlety of my negative thoughts, I am quite surprised. And because it's not getting me what I want, it's obviously time to start deliberately thinking other thoughts. As a start I construct a list of affirmations — a "mental house" in which I can live more productively as I begin work to build one in the physical world.

Affirmations are almost magical in their simplicity and in their effectiveness. If I am what I think about all day long, I need to begin practicing thinking about what I want to be. Affirmations help focus that thinking. Much has been written about the design and use of affirmations, and a book by Joseph Murphy, *The Power of Your Subconscious Mind,*[5] was the first one I ever read. His guidelines were simple:

• The subconscious mind knows the answers to all problems
• Think good, and good follows.
• Know that you can remake yourself by giving a new blueprint to your subconscious mind
• Feed affirmations to the subconscious in the present tense as if they are already true

More than forty printings and a million copies of this book are circulating in the world. There must be something of value there. Using Dr. Murphy's guidelines, I construct the following affirmations:

1. I am able to relax and allow gently, inspired intelligence to operate daily in my life.

2. I am becoming mentally, physically, emotionally and spiritually healthier every day.

3. I recognize problems quickly and find easy, effective solutions.

4. I love my wife and daughter and express that love daily.

5. I draw around me all the good craftsmen, subcontractors and material suppliers I need.

6. I have wonderfully positive relationships with my crew, subs and suppliers.

7. The perfect building parcel is available for me to purchase at a price where I can profit.

8. I am an excellent craftsman and build fine homes.

9. All the resources I need to build a new house are available to me.

10. I easily attract people willing to invest their money and trust in me.

As I begin reciting these affirmations on a daily basis, I notice that very soft thoughts echo the opposite of everything that I say. This is the part of me that doesn't believe, for example, "I am an excellent craftsman and build fine homes." This is the voice that affirmations are intended to activate, for by repeating these affirmations over and over, and by listening to that voice expressing disbelief, I can begin to make the subconscious conscious. I can begin to be aware of what some of the stumbling blocks are. When I know what the blocks are I can begin to make choices that will surmount them. I'll still say the same affirmation, "I am an excellent craftsman and build fine homes," and I'll still hear that same soft voice saying, "No you don't." Only one day I'll be able to laugh it away with the final remark: "Maybe not. But people are paying a million dollars for them!"

Another aspect of affirmations is similar to the familiar phenomenon of noticing things that we have reason to pay attention to. For example, several years ago I bought a white Nissan truck with an extended cab and jump seats where my daughter could sit. No sooner did I drive it off the lot when the road seemed full of white Nissan trucks with extended cabs. I would see two or three a day. When we are paying conscious attention then, to where our affirmations direct our minds, we tend to notice more and more opportunities as they manifest around us. Opportunities, like white Nissan trucks, that we would not have even given a second glance to under normal circumstances.

Researching codes, subs and suppliers starts heading me in the right direction. Working with the affirmations daily, keeps me on the right road. As the readiness grows, so does opportunity. Very soon my discipline pays dividends: a friend of several years, Stan, turns out to be a person willing to invest and trust in me. He has the money to invest and he has already been through the process of building his own new house a year before. I'm astonished to recall that Stan has approached me several times before and expressed an interest in doing a project together. Until now, blocked, with insufficient need, I never took him seriously.

I call Stan and find he's ready to start looking for a suitable property immediately. I'm excited and find a property right away. It's vacant and located near a very busy street in one of the Peninsula's most prestigious (spelled e-x-p-e-n-s-i-v-e) towns. The foundation is shot and the town is considering demanding that the house be torn down. It's on a lot 50 feet wide by 120 feet deep. The asking price, $160,000, works out to right around $30 a square foot once the lot is cleared. This seems high to me, particularly when I consider that a builder has a new house for sale directly across the street that he can't sell for $300,000. When I ask myself why he didn't buy the house I'm interested in, I don't need Albert Einstein to tell me its time to start looking elsewhere.

The next property I find is located on a main street right across from a lake in a section of the peninsula called

Emerald Hills. It's a vacant lot and it turns out that the lot next door, also vacant, will have a house going up as soon as the building permit is issued. I review the plans and talk to the realtor about the next door property and discover that the house is expected to sell for around $375,000. (It's very difficult to incorporate these large money numbers into my head. People act like $375,000 is chicken feed. The first house we built for Bountiful Builders, land included, sold for $42,000 only a few years ago!)

The owners of the lot I am interested in want $100,000 for it. That seems like a somewhat reasonable price. I make a somewhat reasonable counter-offer: $80,000 and we settle on $90,000 contingent upon me being able to get financing. (The typical "Sleazle Clause" that I learned from the Sunday morning ministers, since I already have the financing through Stan if we agree to go ahead with the project.) A real problem does crop up however: the owner wants to retain a 25' wide strip of the property as a second rear driveway to his house located directly behind us. After much analysis and deliberation, Stan and I agree that such a retention is unacceptable and we withdraw our offer. On to the next possibility.

The next prospect shows up virtually in my own back yard. A woman who has migrated from Mexico has purchased a run down old house a block away from me. As soon as she takes possession it seems her entire extended family of cousins, aunts, uncles, nieces and nephews all manage to find their way to this 1200 square foot ramshackle structure. They all manage to be living there at the same time. The foundation on this house in non-existent and a number of repairs have been made without benefit of a building permit. After receiving numerous complaints from the immediate neighbors, the building department condemns the house and requires that it be abated. They also order the utilities turned off. This makes it uncomfortable for the twenty or so people and animals living there, but not so much that any of them decide to look for other living quarters. That ultimately requires the efforts of the city police and a few Immigration and Naturalization officers.

With the house sitting empty the woman who owns it is no longer receiving any rents. Soon she will be in danger of losing the property to foreclosure. I call her finance company to see if they are willing to allow the owner to walk away from the property if it goes into foreclosure and to see if they will sell it to me at a discount, since the house is no longer habitable and no longer securing their interest. I discover that the finance company had foreseen this problem. Not only is the abated house securing their interest, but in addition they have also required the owner to put up her own house as further security. Here, I think to myself is the classic case of a Ted Barkley "Don't Wanter". This woman has to sell this property fast or she will be in jeopardy of losing her own house and whatever equity she has in it as well. What to do?

I make the woman owner of the property an offer of $75,000 through Mike, a friend who owns a small realty company. This is the total sum of the loan still outstanding on the house. My offer will permit her to retain her own house free and clear and she will be out from under the payments that her renters had been making previously. It is a good offer and I am willing to accept whatever karma it produces.

What it produces is a simultaneous offer from another interested party . . . for $80,000. Since Stan and I are unwilling to beat that price, the woman accepts their offer. End of interest.

Not the end of the story, however. Several months later I happen to be in the County Clerk's office researching documents, when I run across a Deed of Trust taken out on that very property. It is a note for $40,000 from the buyer who had overbid me payable to an agent in the same real estate company that had originally listed the lady's property. The listing agent had simply gotten a shill to go in and make an offer $5000 higher than mine in order to buy it for himself! Mike, my realtor, is not surprised when I tell him what I've discovered. "Wait to see how it all turns out," he cautions me. It is with a great deal of satisfaction and a deeper understanding of the law of karma that two years later I witness a new house sitting unsold on that very lot. It sits there unsold for more than thirty months.

Meanwhile, Stan and I continue to look at other properties. A three million dollar abandoned school site. A mountain property looking out onto a large expanse of ugly white greenhouses. Buildable lots in the hills formed by the recent installation of a sewer district. Old, stench-filled wrecks in which people have died and not been discovered for weeks. Houses with second unit possibilities, expansion possibilties and possibility possibilities. All are wrong for one reason or another. I am beginning to profoundly understand the advantages that Dwayne has over me in the perseverance game. Fortunately, I'm doing new remodeling work and the household bills are once again current.

One Sunday afternoon I am on my way to go swimming with my three year old daughter and for no good reason I ride down a little side street in nearby Palo Alto. I happen to drive by a tiny little cottage with a "For Sale" sign in front. There are people inside and it appears they are having an open house. I drive around the block, pull up in front and take my daughter, Amanda, inside for the grand tour.

We find the house is essentially two and a half rooms, very reminiscent of the abated property around the block from my own house. Except this house is extremely neat both inside and out. I come to find out it's a probate sale which will have to be confirmed by the court. The asking price is $180,000 and an offer for $200,000 has already been accepted. Court confirmation will be in five days and the minimum first overbid will have to be $215,750! Almost $16,000 just for being a little late with an offer.

As we are preparing to leave, Amanda stoops to smell the flowers along the walkway.

"Get away from those flowers!" the realtor suddenly screams at her. I pointedly tell the realtor to buzz off, and on the way to the swimming pool, I promise my daughter we'll take care of that lady. We'll come back here with a big machine and knock her whole house down. Forsaking karma for the moment, my daughter and I think this idea smashing.

Stan comes back with me to look at the house and we both decide the lot has building possibilties. The house though, will need to be demolished. A brief review of recent comparable sales in the neighborhood discloses that $450 — $500,000 is not out of the question for a new house, even though it's out beyond anything I can comprehend. A half of a million dollars for a house on a lot measuring 50 x 100! In 99% of the rest of the world you can get whole houses two times the size of this entire lot for less. Still, the numbers at this time, in this part of the world, look feasible. Stan and I decide to take his money to court. Our best guess is that after all expenses are in, we can gross between 40 and 50 thousand dollars profit. We agree that $230,000 is the most we are willing to bid for the property.

Remembering the inappropriate manner in which the listing realtor yelled at my daughter, I see no reason why we should present our offer alone in court and allow that agent to receive the full commission. So I call up Mike once again and ask him to present our offer. He agrees and receives half the commission when our overbid of $215,750 goes uncontested. Unlike the flower-loving realtor, Mike likes Amanda quite a lot.

With the transaction completed, my work as a spec-builder is begun. In less than three months I have gone from being virtually broke and out of work, to committing myself to a project that will probably take a year to complete and involve hundreds of thousands of dollars. I'm both excited and apprehensive about the lessons laying in wait. But I'm ready and willing to learn and grow.

3

"Man alone is the architect of his destiny.
The greatest revolution in our generation,
is that human beings, by changing the
inner attitudes of their minds, can change
the outer aspects of their lives."
— William James

Drafting the Plans

Once we obtain title to the property, Stan and I begin discussions on how best to go about getting the house designed and permitted. He has heard encouraging words about a local architect named Toby Delhi, whose office is only a few blocks from our building site. It matters little to me who the architect is and I agree to have Stan and Toby work out the design and engineering requirements for New House. We have purchased the house in May. Toby and his assistant, Lou Anne, assure us that they can have plans ready for review by the Palo Alto building department by July.

In the meantime there is much to do. I must get a crew together, keep work scheduled up until the time the house construction is to begin and decide how best to demolish the existing house — whether to rent debris boxes and take it down slowly piece by piece with day-laborers as fill-in work, or simply hire a demolition company and bulldoze it all at once. A check with the building department makes the decision very easy: no demolition permit will be issued until the plans for New House are completed, submitted and approved. The old house must remain until the city okays a new one to take its place.

I have a number of my own concerns with the design of New House. I've worked long and wrong often enough to know that I will not be happy to build a great house and lose money on it. Thus one important factor is that I want New House to be profitable.

I also want it to be easy to build. Some of the factors that will contribute to ease are: a roof with no greater than a 1/4 pitch (6 in 12 or 26 ½ degrees), a stucco exterior, no overhangs, drywall interior window trim, and carpeting over plywood throughout except for the kitchens and baths. However, Palo Alto City Ordinances impose pretty severe restrictions on what can and cannot be built on our new lot. Many factors have to be considered simultaneously. New House can be 60 feet long — which isn't a problem — but it can only be a maximum of 38' wide and no higher than two stories.

Stan works regularly with Lou Anne, who has primary responsibility for our project. I have made my preferences known to both of them. Given the lot restrictions, the first preliminary floorplan has made the best possible use of the buildable space. Here is how it shapes up (*see Illustration 3.1*).

Stan and I are pleased with this design and we can begin to get a sense of how the house will look on the lot. We approve this preliminary floorplan and several weeks later we meet to take a look at elevations.

The elevations are not at all what I had in mind. Half the roof is gabled (shaped like an A) and the other half of the roof is flat. (*See Illustration 3.2*) I abhor flat roofs. They pond, leak frequently, and can never be adequately repaired for any great length of time. Even five-ply, built-up tar and gravel roofs need to be replaced before a pitched roof will. I also abhor having to work on steep roofs. This design has a 1/2 pitch roof (12 in 12), essentially a 90 degree angle, which is very steep. Steep roofs and flat roofs taken together leave me flat and cold. Not only that, but Toby and Lou Anne have specified no stucco. Instead, 6" wood siding is to go all around the exterior of New House.

This is not turning into the house I wanted to build, however it is a joint project. Because I have turned the responsibility over to Stan, I don't have much room to protest. Besides, he likes the design, and truthfully I haven't been able to come up with a roof design that looks better. I've tried flattening it, lowering the pitch, turning it and nothing seems to really work. A mansard roof, with the lower part pitched like a leaning wall and the top almost flat, is the only thing that looks half-way decent, and I am not all that crazy about mansards. They are still essentially flat roofs.

Illustration 3.1

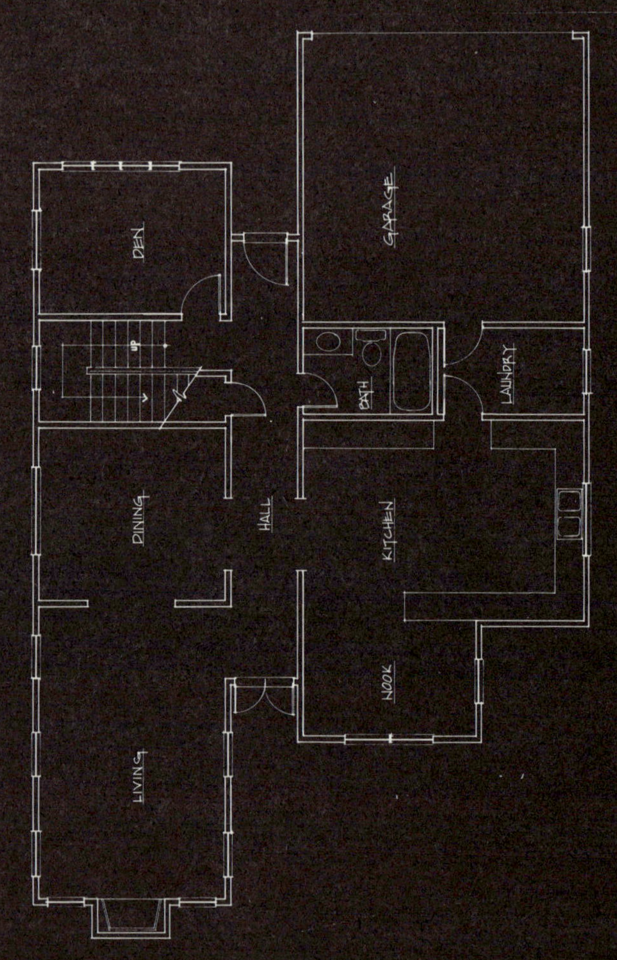

DEN
GARAGE
UP
DINING
HALL
BATH
LAUNDRY
KITCHEN
NOOK
LIVING
FIRST FLOOR PLAN

Illustration 3.2

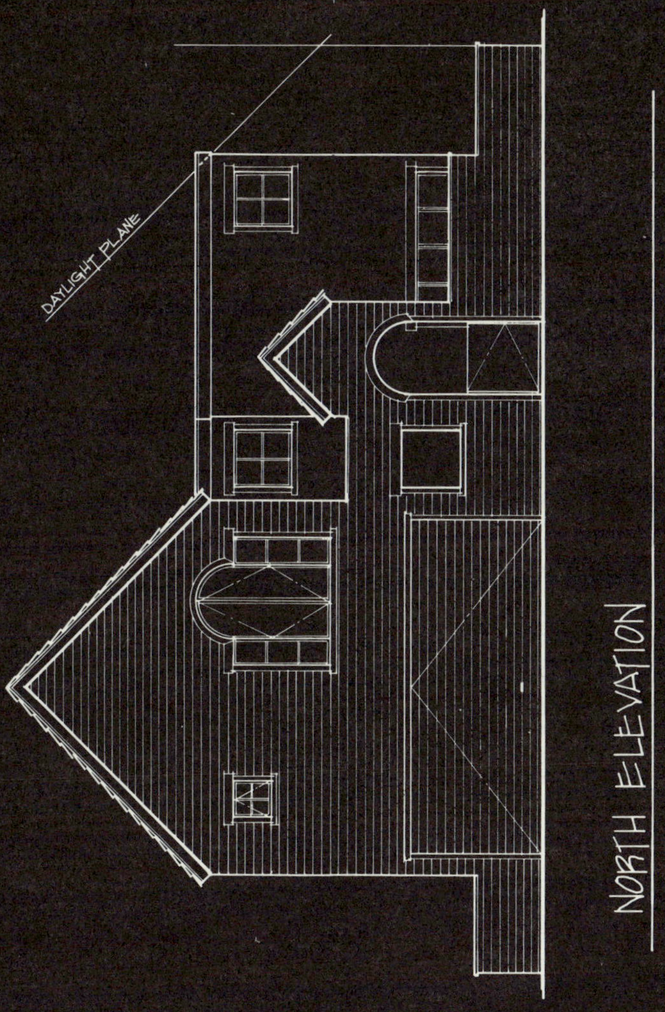

Through the course of much discussion and deliberation I happen across a short story that helps me gain some perspective. It's called *The Illusionless Man* by Alan Wheelis. It begins

Once upon a time there was a man who had no illusions about anything. While still in the crib he had learned that his mother was not always kind; at two he had given up fairies; witches and hobgoblins disappeared from his world at three; at four he knew that rabbits at Easter lay no eggs; and at five on a cold night in December, with a bitter little smile, he said good-bye to Santa Claus. At six when he started school illusions flew from his life like feathers in a windstorm; he discovered that his father was not always brave or even honest, that presidents are little men, that the Queen of England goes to the bathroom just like everybody else . . . Happiness was of course a myth; love a fleeting attachment, a dream of enduring selflessness glued onto the instinct of a rabbit. At twelve he dispatched into the night sky his last unheard prayer. As a young man he realized that the most generous act is self-serving, the most disinterested inquiry serves interest; that lies are told by printed words, even by words carved in stone; that art begins with a small "a" like everything else . . . He became a carpenter but could see a house begin to decay in the course of building — a perfect pyramid of white sand spreading out irretrievably in the grass, bricks chipping, doors sticking, the first tone of gray appearing on white lumber, the first film of rust on bright nails, the first leaf falling in the shining gutter

. . . in all that he did he could see himself striving toward a condition of beauty or truth or goodness or love that did not exist, but whereas earlier in his life he had always said, "It's an illusion," and turned away, now he said, "There isn't anything else," and stayed with it.[1]

With regard to the design for New House, I too must say, "There isn't anything else", and work to let go of how I think it should be, while simultaneously resolving to stay with it.

It is important to have a reasonably solid sense of one's strengths and weaknesses. After considerable experience, I've concluded that construction work and design work require two very different mental and emotional sets. With so much of the day-to-day emphasis in my life on construction details and problem-solving, most of my attempts at design have been singularly unimpressive. Consequently, when I'm able to let go and leave the designing to Lou Anne, Stan and Toby, I feel this serves everyone's best interests. My focus shall remain on the actual building of New House.

In early September the plan is complete and ready to be submitted for the city's approval. Half the roof will remain pitched steeply and half will be flat. The siding will be 6" red cedar. Overall, given the lot restrictions, the resulting design really does best serve New House and the construction process. Because the lot is long and narrow, the house must be tall and narrow to obtain a sufficient amount of square footage. This will give the home the look of a townhouse, which is not unfortunate or out of character for the neighborhood. Wood siding will give detail and visual relief to the large, wide, front windowless areas, whereas stucco would not. Also, siding can be applied to the garage door to allow us to blend it in and draw attention away from the door as a prominent feature at the front of the house.

As they go to the Building Department for approval, the front exterior elevation of New House looks like Illustration 3.2.

In mid-October the plan is back to us for reworking. The plan checker doesn't agree with our energy conservation calculations, which are required for all new houses by the State of California, nor does he like our lot coverage ratios, or the sizes of some of our beams. To appease him we have the engineer re-do the calculations. We also upgrade all exterior doors to double glazing, increase the roofing and attic insulation and reduce the amount of footprint on the lot by pulling the dining room wall in three feet. These changes made, the P. A. plan checker is appeased.

The City of Palo Alto has a Daylight Plane Ordinance, or DPO*, which means essentially, that you are not allowed to block all of your neighbor's daylight. You are allowed to block 12 feet of light at the property line and 18 feet of it when you are set back six feet from that line. Because our lot is so narrow we have taken New House out to the maximum on both sides — a somewhat risky approach. Perhaps a better idea would be to allow a six inch margin on each side. Then we won't have to locate the house on the lot so precisely. Also, the extra margin would afford some leeway with regard to our height in the daylight plane.

But no decision for changing is made and so we keep the house dimensions the same. However, we do determine that the wooden first floor joists on a concrete perimeter footing will raise the total height of the building to a point where it will significantly intrude into the daylight plane. The most elegant solution is to simply support the house on a slab. We decide to revise the plans and resubmit them for final approval with all the requisite changes together with new plans for a concrete slab foundation. In mid-October the plans are finally returned to us along with our permit, and the building process is ready to begin.

That evening I have the following conversation with my wife, Susan:

"Tomorrow we will start demolition and begin New House."

* *"The daylight plane shall be an inclined plane, beginning at a stated height (12') above average grade, that average grade being an average of the grade at the midpoints of each side of the property line and between the grade of the site and abutting sites, and extending into the site at a stated upward angle (45 degrees) to the horizontal, which may limit the height or horizontal extent of structures at any specific point on the site where the daylight plane is more restrictive than the height limit or the minimum yard applicable at such point on the site."*

"You seem excited."

"A little. Excited . . . and scared. I mean it's a whole new house. A dream beginning to materialize, but also months of work, and decisions and planning. Months of mistakes."

"You're planning to make mistakes?"

"I'm not planning to, but mistakes will inevitably be made. Part of my job is to recognize when a mistake has been made and to deal with it directly and expediently. I will have to watch the tendency to deny and avoid. Denial and avoidance always have dire consequences further down the line."

"Sort of like a do-it-yourself karma factory."

"Right. Cause and effect are laid out in very close correlation in house construction. Crooked walls and leaky joints make themselves apparent all along the way and remain apparent every day you have to spend on the job. I can walk into every house I've ever built and immediately point out every single flaw and error that went undetected until it was too late to make corrections. It's only as I've gotten older that I am learning to simply observe the flaws, accept that they will be there, and instead focus my appreciation on the larger accomplishment. The richer feel and flavor of the complete achievement."

"It's good that you bring *only* that talent for observation and appreciation into our marriage . . ."

4

"If you have built your castles in the air,
your work need not be lost;
that is where they should be. Now put the
foundations under them."
— H.D. Thoreau

Laying the Foundation

Every large work requiring sustained and dedicated effort is usually accomplished in cooperation with supporters and collaborators. For the lumber-mechanics of New House, my principal collaborators and supporters are Mitch, Ramon, Carlos, and of course, Stan.

Mitch is a licensed general contractor who was introduced to me by mutual friends before New House was even in the pre-planning stage. Mitch is a competent craftsman, an amiable collaborator and a dependable worker. He's the man I rely upon most to take care of business when I am away from the job. He's 32, athletic, and his wife thinks he has funny hair.

Ramon is an immigrant from Uruguay. I have been training and working with him for almost two years as New House gets under way. Ramon is only 22 and, prior to coming to the U.S., he worked in Uruguay as an electrician's helper. He is very strong and very smart, having learned competent English in less than six months. He also has a work permit, a good sense of humor and, except for the fact that I have to pay him extra to come to work on time each morning, he's joyous to work with.

Carlos is another immigrant. He escaped from a prison in El Salvador and fled to the U. S. when he was 18. He's now 19, has no building experience and speaks virtually no English. Carlos is small and unreliable — in part because of the language problem, and in part because no one has ever asked him to be otherwise. He's willing to work cheap, but I pay him the same as I would any other high school teenager. He works best with constant supervision, but on New House he must be supervised by Ramon, the only one of the crew who speaks Spanish. It doesn't make rational sense to have Carlos working with us on New House, but as I rediscover over and over, many aspects of this job, and many aspects of the whole rest of my life, don't always make rational sense. Anyway, Carlos has a great smile and, when he calls me "Senor Marco", he makes me think one day it might be possible to head the entire Northern California Spanish Underground.

Stan is the financial and co-creative thrust behind the

New House. We met several years before through work our wives were doing (counseling people who are either dying or grieving). I liked Stan immediately. He's smart, successful, and skinny. Like many people who go into business with the intention of providing essential goods and services to mankind, Stan has enjoyed a life of prosperity and material comfort. He's also corny and looks enough like me that his friends sometimes mistake us for brothers. His strengths dovetail very closely in areas where I feel I have significant deficits, for example, with financial analysis. Our spoken objectives are to design and build a fine house, have fun learning and working together, and to make a reasonable profit on the value we add to the property.

Finally, there is me. As I chronicle the adventures of me over the course of the year it takes to complete New House, different fragments of my personality appear on the job in a number of different guises. Many of those fragmented aspects will be presented and described as they find their way into and through the building of New House.

Several weeks before the architectural drawings for New House were ready to go to the engineer for him to do his work, I gathered up several sets of the foundation plan and sent them out to a half dozen concrete contractors to obtain bids for putting in the slab. All the bids come back with less than a 10% variation. The job then becomes how to discriminate between them.

I immediately eliminate two contractors who indicate that they cannot guarantee they will be available within a week after we obtain our permit. I eliminate two more who want me to pay half the money up front for materials. They are too poorly capitalized and their credit is suspect. Of the remaining two, discussions with several contractor friends reveal that one of them does extensive work for the City of Palo Alto. They are a family business: a father and four sons with a few cousins thrown in for good measure. I like Robert, the field contact, for his ready smile and friendly knowledge. I ask him to send me a contract which I sign and return, promising to contact him as soon as the permit is issued and the lot is staked, cleared and compacted.

Establishing clear property lines is another essential task required before the actual construction work can begin. Fences already surround three sides of the property and it can be safely assumed that the front line begins exactly two feet behind existing the sidewalk. The depth of the property works out precisely from that two foot measurement, but the sides come up over a foot short. I have been told that the first question I will be asked when the foundation forms are inspected is how I determined the property lines. So I contact a surveyor who has done other work directly around the block and ask him for a bid. He quotes a price range of $700-$1100 and indicates that, because of his recent work nearby, it will probably end up nearer the lower end of the range. I tell him to come out and set the stakes.

Two days later the survey crew drives up at approximately noon. This crew consists of two guys, only one of whom gets out of the van. He goes digging around each corner of the property, then walks back to the van.

"Found them all," he reports to the driver. He then

measures distances between each located iron pipe, pounds in four wooden stakes, paints the pipes and stakes orange and they drive away. It's one o'clock in the afternoon.

"Gee," I think to myself. "$350 per man per hour. Not bad."

The next day I call their office, unable to believe they really intend to bill me at that inflated rate. They weren't — they were intending to bill me more — $750 dollars to be exact! By their calculations they were billing for 6 hours of work. I just saw the field work, the tip of the iceberg, they explained. I didn't see the laborious office work. Clearly, the office workers told them that the field work would be so difficult that the actual worker would be so exhausted he would need a chauffeur, which is why they sent two guys. I resign myself the situation and accept their estimate. What's done is done.

Three days later the surveyor's bill arrives for $1100!

Angry and determined, I call the office manager immediately. Coincidently, he is on vacation. So I talk to the billing woman and tell her I am only paying $750.

I do pay that, but they keep sending me bills which I ignore. When the office manager returns from vacation and sends me a nasty note, I decide to contact the owner of the firm directly. I tell him he was recommended by a mutual friend and I'm not happy with the treatment I'm receiving from his manager. I carefully explain my position, tell him what I want and listen to what he has to say.

Several days later the owner calls back and tells me the office manager has just been let go for many of the very reasons about which I complained. Also, he will see that my billing is corrected immediately. There is some justice in the world and there is some learning for me. On the next house I won't end up paying two guys $375 an hour for clearing some weeds and exposing boundary markers. Now that I know enough to do it, I no longer need a surveyor's experience and authority to explain how I have determined the property's boundary lines.

I have two estimates to clear our building lot. One for $2500 and one for $4500. I ask the lowest bidder just what he is going to do for the money. It's the exact same work as the high bidder: take away the building, remove all concrete, asphalt, trees, and shrubs and clean up all resulting debris — in essence create an empty dirt lot.

The permit is issued in the second week of October and demolition is slated to begin on a Monday morning. I have promised to take my daughter to witness the destruction of the house where the lady yelled at her for smelling the flowers. She is very excited to see this.

When I drive up to the lot however, I am confronted with a virtual traffic jam, and half of the cars seem to belong to the Palo Alto Police Department! It seems that every single one of my new neighbors has called the police, taking exception to the 7 AM starting time that the demolition company has decided to give themselves. This is in direct opposition to my instructions to begin at 8, and it gets our work off to a difficult start in the neighborhood.

I then discover one of the company's workmen has taken exception when the Palo Alto Police ask him for identification. When he eventually is persuaded to produce some I. D. by four other police officers who join the party, they discover he has a current arrest warrant out-standing. So the house will be demolished one man short.

My daughter, Amanda, loves the excitement. She thinks policemen are her friends and that they look dashing in their uniforms. She also thinks they always catch the bad guys.

After the police leave we watch the bulldozer take big bites out of the house, load them into dump trucks and cart them away. In the span of an hour the house is com-pletely gone. As we walk along the edge of the property back to the truck, Amanda bends down and comes up holding an old rusty horseshoe. She gives it to me and we hang it on a nail on the fence near the front. As we continue walking, one of the demolition men comes over with a hammer and another nail. He takes the horseshoe down and renails it back with its ends pointing upward.

"That's so you can catch the good luck," he tells us. "Put it over the door legs-end up once it's built. Then you can keep the good luck coming." Little do I know just how much I'll need it. But for now, we are happy: the house was demolished, as I promised many months be-fore, and we found a lucky horseshoe. Amanda goes off to nursery school with her day made completely.

Switching over to a slab foundation presents us with serious challenges. Heating ducts will have to be placed high in the walls of the first floor or else located precisely in the proper area prior to pouring the concrete. This is something that is rarely done accurately or without the ducts getting stepped on and blocked during the frenzy of the pour. In New House they will be placed overhead.

A similar problem exists with the plumbing. With a raised wooden foundation that has a crawl space below, drain and waste vent pipes can be placed precisely after the walls are framed or laid out. With a slab everything has to be measured from imaginary walls (which never end up exactly where originally imagined), and much care has to be taken that pipes aren't broken and joints loosened. I've seen enough slabs to know it wouldn't be the first time a main vent stack has had a cap knocked off and half a foot of wet concrete inadvertently poured into it.

Originally, I expected that New House would be started in August and framing would be well under way by Thanksgiving. Given this schedule, my wife and I made plans with several friends to rent a cabin and go away for five days over the holiday weekend. Now that vacation time has arrived, the slab foundation is not even trenched or formed. I talk it over with Stan and we agree that I should go anyway and leave Mitch, who's most experienced, in charge. In an emergency he can contact me by phone.

No sooner do I arrive at the cabin than Mitch is on the line: the forms for the foundation have not passed inspection. Inspector Phillipe, one of the Palo Alto Building Officials, contends that the footing depths are incorrect. The plans call for them to be 26 inches deep and they are only 24. Also, the widths in the garage are supposed to be 8 inches and they are only 6. A call to Robert, the concrete contractor, slowly brings the problem to light. The original plan I gave him for preparing his estimate was later revised by the engineer who added both those two inches. Neither Robert nor I noticed the changes he had made. The first mistakes are already beginning to surface.

Since I consider it my mistake, I instruct Mitch to work with Ramon and Carlos, hand dig the remaining two inches, and reset the forms in the garage to the required 8 inches. Then reschedule another inspection and call me with the results. I am encouraged when the day after Thanksgiving Mitch calls to tell me that everything has been completed and Inspector Phillipe has signed off our foundation forms.

While I should be happy at this news, my wife recognizes my feelings are mixed.

"So how does it feel to know that you really aren't necessary to get this project going?"

"Good, actually. And bad."

"Good how?"

"Good that I can relax and have confidence the work can be done without having me on hand to supervise every detail. Good also that I don't have to always be available to take the heat from the authorities. I'm not always up for such heroics."

"How bad?"

"I feel bad knowing I'm really not needed at all. I'm a dispensible commodity — an easily replaceable entity. I wouldn't be surprised if much of the work is actually best done with me not there."

"You wouldn't be surprised? Would you be grateful?"

"Probably not."

"Maybe you just need a little more practice. Here. You might enjoy this book. Especially chapter 5."

The chapter that my wife suggests to me is from a book entitled *Spiritual Economics* by Eric Butterworth. I'm not big on organized, large-scale religion and this book is published by The Unity School of Christianity. The chapter she suggests is called "The Grateful Heart". I figure it is Thanksgiving so, what the hell. While the rest of the group has gone off to town, I take time out to read it.

"My most important asset is the conscious control of my own life."

Interesting thought. I would probably say: *"An* important asset", and I would also suggest that "potential control" is more in line with my own daily reality.

"Nothing else will satisfy or fulfill unless I enjoy the freedom that comes from control of my inner world of mind and emotions."

My experience says this is true, that it is not easy, and takes a lot of practice to accomplish. I am reminded of an exercise suggested by Robert Fritz, a Boston-area musician. He calls it The Pivotal Technique[1] and it's pretty simple. Step One: Assess where you are, what you are doing, or what you have. Observe and note it. Step Two: Determine where you want to go, what you want to do, or what you want to have. Note that. Then note the degree of disparity between where you are and where you want to go, or between what you are doing and what you want to do. End of formal exercise. However, personal next steps will show up in quick and sharp relief.

I return to reading the chapter on "The Grateful Heart".

"You are not obligated to thank God for your life, for your work, for your prosperity. However, giving thanks is an important state of consciousness which keeps you in awareness of the universal flow of energy. When you understand this you see that a grateful heart does not need something to be grateful for. One can be grateful with the same spontaneity as being happy. It simply flows forth from within and becomes a causative energy."[2]

As I read this passage I recognize what the experience of feeling grateful feels like. I can readily connect it to people, places and possessions, but I also recognize that it has an energy all its own. There's a peaceful quality about gratefulness — a calming sense that everything is perfectly splendid just the way it is. Unfortunately, it's an awareness often far removed from the daily me-mechanics and the lumber-mechanics of house building.

"A grateful mind is a great mind which eventually attracts to itself great things." — Plato

"The grateful heart will always attract to itself in one way or another, through human hands, or through wonder-working ways, the great things needed to solve a particular situation. It is an outworking you can stake your life on."[3]

I get the impression from Butterworth and Plato both, that my heart and mind are like great big problem-solving magnets. By simply being aware of and expressing feelings of gratefulness all solutions are made manifest. It almost seems too easy and, at the same time, too hard. How do I walk around feeling grateful all the time? How do I go about expressing that gratitude? I'm afraid to read on, lest I find out.

Precisely at that moment my wife and friends return and I give this report:

"Guess what. I found out that Thanksgiving is a good time to practice giving thanks. It's also a good time to get foundations inspected by hardnosed building inspectors. I've also found out that its time for me to change my name: from now on friends, call me . . . Captain Grateful. You, my wife, can call me "Your Holy Gratefulness".

After this pronouncement I am simultaneously regaled with a thundering round of asymphonic Bronx cheers. Each to their own ways of giving thanks, I guess.

5

"That was a happy day, before the days of
architects, before the days of builders."
— Seneca

Under Critical Eyes

Shortly before the building permit is scheduled to be issued, I send a letter to every neighbor within eyeshot and earshot of New House's construction site. In that letter I introduce Stan and myself and tell them we will be tearing down the small house and building a new one in its place. I send along copies of the elevation plans and ask for their input. I also ask people to bear with the noise of construction for the next few months, and if they have any concerns to please get in touch with us directly.

Only three people ever make contact. The neighbor

directly across the street (who is renting) walks over to say that she appreciates receiving the letter and knowing whom to contact with her concerns. She doesn't have any at the moment and hopes our project will be successful. She also thinks our front elevation facing her house will be a pleasant change.

Another contact is made by Homer, the neighbor to the north side of New House. Homer calls to inform me that, in the course of demolition, the earth movers have caused his fence to lean somewhat (it was already leaning under the weight of his overgrown shrubbery) and he'd like me to fix it before it falls over (it won't) and his dog loses his play area. I see no point in telling him that it is not his fence, as it is located almost a foot on our side of the property line and runs the full length of our lot, a good fifty feet beyond his corner property. Instead, I thank him for calling and tell him I'll secure the fence in the morning. Before he hangs up he lets me know that he hopes I'll be putting in a new fence, but that I shouldn't expect him to lay out any money for construction costs. Thank you, Homer.

The nightmare is just beginning over on Homer's side of the property. Later, when we are well along in construction, Homer informs me that I now owe him a new fence for all the noise I've been making and for all the inconvenience I've caused him by having our cars parked on the street. He also lets me know that he knows I'll be making a fortune with New House and that I can therefore afford to pay for the whole fence.

Wishing to bring peace to my relationship with Homer as quickly as possible, I go ahead and build the fence between our properties. The morning after the fence is finished, I drive up to the job and find Homer outside with his measuring tape.

"We have a problem," he informs me. "This fence is not built where it should be. It's too far over on my side."

"Really, Homer? How far over is it?" I ask.

"From an inch and a half to two and a half inches."

"That much, huh, Homer?" I feel the heat rising. I want to tell him what I think, that he's a self-absorbed, ungrateful, stingy, conniving earthworm, but I resist the urge.

"Not only that, but it's too high in front of my house. I can't see down the road now like I could before."

"It's built to code, Homer. Look it up. Four feet in the front and six feet at the setback."

"I don't care about the codes. I want it lower and I want it moved over. I'm a lawyer and unless you want to spend a lot of money and unproductive time in court, you better fix it."

Standing before Homer now, I am livid.

"I'll talk it over with Stan and I'll get back to you," I tell him. Then I turn and walk away, imagining a scene from Badlands where Marlon Brando turns and hurls a razor-sharp spiked ball through an adversary's neck. I think the adversary there was a lawyer too, if memory serves.

Stan and I do talk about Homer. First we fantacize all the ways we can make his life more miserable than it already must be: We'll find some old auto wrecks and epoxy them to the road right out in front of his living room window. Then we'll cut an intermittent short that can never be found in his TV cable. Next we'll coat his mangy dog with asphalt cement so he can track it all over the walls, carpet and linoleum of Homer's house. After this, we feel much better and decide to lower the fence a bit all along the front, but decide not to move it any further over onto our side of the property. When Homer comes to complain, we point out the surveyor's stake that cost us so much to locate, and show him how the fence exactly straddles the property line. For the time being he goes away . . . mad.

The first contact with Brenda and Larry, the neighbors to the south, begins on a less toxic note. Larry is a well-known gynecologist in the community, a workaholic and an active campaigner for aids and world peace issues. Brenda, a raven-haired beauty with big brown eyes, is a psychiatric social worker on sabbatical while she raises their three children. She greets me on the site one day and asks if she can have the stone bird bath sitting abandoned in the front yard. I tell her she's more than welcome to it and ask if she'd be interested in splitting the cost of a new fence between our property. She agrees right away and is also willing to pay us extra to extend it the full length of her property (which is deeper than ours). My conversation with Brenda is a relief after Homer. I feel like I'm dealing with a rational person. I casually ask her what she thinks of our plan for New House. In my mind this is simply a conversation-making question. In her mind, it's an opening to launch a new career as an architectural critic.

"Don't you think its a little big for the lot?" she asks. "And I really think it should be moved farther from our side and set back more from the front. Personally, if it was my house, I would build it as a single story."

As I stand listening to this impromptu lecture in Architectural Criticism, I am looking directly at her house. It is huge and almost completely fills her lot. It's also closer to our property than the six feet current law allows (probably grandfathered in as a legal, non-conforming structure). It's a single story house, but it has a basement and it sits up on a foundation a good six feet higher than any of the other houses in the neighborhood, including ours. New House is located as far front as the code permits simply to allow us to have a reasonable back yard — and this does make it sit farther forward than hers.

"You seem to have given a lot of thought to this house going in here," is all I say in response.

"You bet I have. And I think your architect is a real jerk! I mean, did he even come out to look at the neighborhood? This house just doesn't fit here. (Her house is pink and purple). I'm sorry we didn't buy the lot ourselves."

"How come you didn't?"

"We couldn't afford it. We just sold our house on two acres in the country to move here into the city and we're having plans drawn up to put an $80,000 pool in the backyard. Things are so expensive. There was just no way."

"Well, I'll be sure to tell your concerns to my partner and to Toby and Lou Anne, the architects."

"Good. Do that. I mean, this house is going to be blocking all of my light!"

When Brenda leaves I'm wondering what I did to deserve this. Then I remember — I wrote a letter asking the neighbors for feedback. Got to be more careful about asking for things. When I find myself wishing I had turned the hose on her and spanked her with a 2x4, I realize that what I really wanted was praise. I certainly didn't want a confrontation with a narcisstic hysteric who sees herself victimized by coldhearted, greedy developers. If light is so important, why didn't she stay out in the country?

Richard Bolles, a consultant who helps people find work that suits their nature, has described the mentality of people who consider themselves victims:

The Victim Mentality

The Victim Mentality, simply defined, is that outlook or attitude which says: 'My life is essentially at the mercy of vast powerful forces out there and beyond my control. Therefore I am at the mercy of —

☐ My history, my upbringing, my genes, or my heritage.

☐ My social class, my lack of education, or my lack of intelligence.

☐ My parents, my teachers, or an invalid relative.

☐ My mate, my partner, my husband or my wife.

☐ My boss, my supervisor, my manager, or my co-workers.

☐ The economy, the times we live in, the social structure, or our form of government.

☐ The politicians, the large corporations, or the rich.

☐ Some particular enemy, an irate creditor, God or the Devil.

☐ Next door neighbors named Brenda and Homer.[1]

The difficult question for me is, how do I deal with people who are unconsciously feeling vicitmized and powerless, who then attempt to victimize and control me? Like Homer and Brenda, for example.

Presently, the best I know how to do is refuse to engage. I don't respond to any of her attacks or criticisms, even though I do feel attacked and criticized. Brenda draws out those angry feelings that make me want to hose her down and beat her with a stick . . . for her own good, of course.

Richard Alpert, formerly known as Ram Dass, suggests an alternative role in which to cast Brenda and Homer — that of teacher:

> "The (teacher) is anybody or anything along the way that helps you along the path a little bit. So even your enemies are often your teachers because they wake you up to the place you are not, which helps you to get free of that place, which helps you to get on with life . . . you work on your own consciousness . . . and you learn to honor everybody you meet as your teacher, for you see that there is nothing else you can do but be conscious."[2]

When I was an undergraduate in college, I often did more research on my teachers than I did for the courses they taught. My sense was there were teachers that were right for what I wanted, and was ready and able to learn, and other teachers that were wrong. My wife is a right teacher for me. Brenda is wrong, wrong, wrong. She would drive me over the brink. I'd have too much to learn and my life would be too short a time to learn it. Still, Brenda lives right next door to where I am working five days a week. How do I keep from victimizing or being victimized by the wrong teachers?

For starters, I resolve to simply watch what goes on. I listen to what she's saying and watch how I'm reacting. I assume that when I feel upset, something other than her words is causing me to feel that way. I usually don't know what it is, so watching my feelings becomes a bit like a detective game.

I begin by asking "Who does she remind me of?" "What other times have I felt this way?"

It's usually childhood stuff. Parent or sibling stuff. Once I identify and connect the roots of the feelings, I'm usually less bothered by what's happening in the present. This does't mean that Brenda is no longer a tiresome pain in the ass. It just means I have more choices available to me in deciding how I want to respond. Hosing her down now becomes a more conscious choice. Calculated and pre-meditated; a more consciously gratifying prospect.

Through all this conflict and stress I still somehow manage to get the fences built and the foundation poured and finished. The further teachings and growth that building New House has to offer lie earnestly in wait, up ahead.

6

"Something there is that
doesn't love a wall."
— Robert Frost

Setting the Framework

Once the slab is cured, the framing lumber is delivered. It is the first Monday in December. We have one month to get the roof on before the January rains make life damp and dreary. But first we need to frame the walls and the second floor.

There is much I am grateful for and appreciate at the framing stage of a house. I love the smell of the new fir early in the morning. It is hard to imagine that a group of minds and bodies will rearrange the scattered stacks of materials into an organized, cohesive whole that is a house.

I also like the feel of the lumber and the movements required to put it in place and nail it all together. Through a whole series of small completions, one wall after another gets laid out, nailed together and braced into place. I relish the initial speed with which the house quickly begins to take shape. In less than three days the whole first floor framing is up and in place. Passersby remark on our rapid progress and wonder how so few workers can accomplish so much.

On Friday, during lunch, Mitch poses the following question:

"What do you think is the fastest time anybody has ever built a house?"

I can feel the trick in the question itself. "Two hours." I answer off the top of my head.

"Close," is Mitch's reply. "Two and three quarter hours — including slab, finish painting, and landscaping!" He hands me a recent clipping from the Wall Street Journal:

"House Constructed in Few Hours Isn't Home Sweet Home"
By Kathleen A. Hughes

San Diego, CA — A starter gun is fired and a herd of 700 burly construction workers, divided into Team A and Team B, race toward two yards. Their mission: break a four-hour-and-18-minute record for speedy home building and set a world record.

An unusually fast-drying cement is poured, and within 45 minutes the teams are running with walls toward the concrete slabs. Building inspectors dressed as referees are in hot pursuit, and the announcer explains excitedly that these aren't prefabricated homes — the teams are building the nine-room homes from scratch.

Not without incident. "There's a problem, quite honestly, with B house," says sportscaster Ted Leitner gravely. "The main roof is on a little crooked." That gives Team A the lead. The A's wrap up their house, complete with landscaping, after two hours and 45 minutes! A chant goes up: "We want beer! We want beer!"

Charles and Angelyne Van Gaasbeck are sitting in their living room watching this scene on videotape with glum expressions. They bought the losing B team's house for $89,000.

"We want beer. We want beer," mutters Mrs. Van Gaasbeck as she crochets.

The event was a 1983 publicity stunt designed to draw some glory for the depressed San Diego building industry. It was preceded by the building of two "rehearsal" homes on the same block, put up by Teams A and B respectively in about six hours each. The problem: The Van Gaasbecks and the buyers of the two rehearsal homes say they weren't told about the contest before they agreed to buy the homes the following year for prices ranging from $89,000 to $93,500.

And they say that life in the homes built for glory has been a nightmare. The single-story homes have cracked slabs, lumpy floors, bowed walls, rocking refrigerators and cabinet doors that won't close — to name just a few of the problems . . .

While they did notice some flaws, the Van Gassbecks and the buyers of the rehearsal homes say they bought the houses anyway because they came with 10-year warrantees. But now they say that the problems have grown worse and that many of the needed repairs haven't been made despite long battles with the seller and the warranty insurer . . .

The Building Industry Association of San Diego County thinks the fuss is unwarranted. "Those homes are structurally more sound than 90% of the homes in San Diego County," says Robert Morris, an executive vice president of the association . . .

"Great idea," I remark in response to the article. "It's too bad they built a bad house. But a bad house doesn't have to be the result just because they built it in 2 ¾ hours."

"Heck, those are probably the same houses that these guys are putting up in tracts that normally take four months," Mitch says. "They only built them badly because they think that's how they're supposed to be built!"

New House will take considerably longer. Six months, weather permitting, is what we have planned.

At New House, the framing of the first floor walls discloses several problems that were not anticipated in the designing stage. The first is that the cathedral ceiling in the breakfast nook is simply too high in proportion to the size of the room. Second, an unneeded, framed closet intrudes into the kitchen and takes up valuable space.

The closet problem is easy. Stan and I simply agree to

take it out. The breakfast nook is not so easy. We ask the architects, to come over and take a look. They propose we drop the ceiling down two feet and frame a false gable inside the original so as to preserve the exterior proportions (*see Illustration 6.1*).

I don't like the idea of having to frame in the small space formed between the original gable (already in place) and the two foot envelope that will be newly created. Instead I propose that we frame it flat at nine feet and make a secret storeroom above (I envision access through a bookcase hooked up to an automatic gate opener that works by remote control). None of the others are so excited about a secret room, but they do agree that we might as well make use of the space. We'll make it offer some utility by framing it into a study above, and a breakfast nook below.

As framing for the first floor progresses, two more challenges surface. First, I have misread the plans and placed a horizontal beam across the middle of the hallway at the first floor level. The beam was intended to go one floor up and become buried in the ceiling. It has to be moved, but it is a difficult task because it is tied together deep into the walls and the floor on each end. I turn this job over to Carlos and Ramon and, though they will spend the better part of the morning on the task, I feel relieved that I can confidently delegate this task and not have to worry about it further.

The second challenge concerns the three exterior walls that make up the living room. These walls are designed to be constructed at a height of 12 feet. Because I have forgotten to include 12 foot studs in my lumber order, I

Illustration 6.1

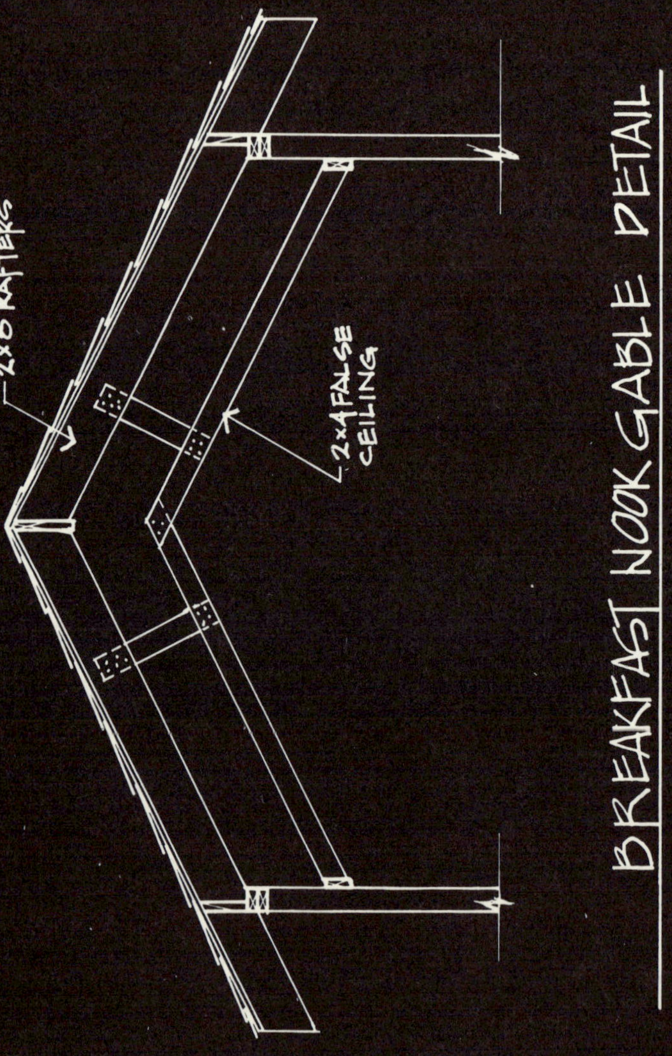

2×6 RAFTERS

2×4 FALSE CEILING

BREAKFAST NOOK GABLE DETAIL

have built these walls at 8 feet and am planning to put a four foot curtain wall on top of them. My expectation is that the roof bearing down from above will only be a vertical load. Therefore, the plywood nailed over the joint on the outside and the sheetrock installed over the joint on the inside will keep the walls from folding in the middle like a sandwich. Still, I don't like the mistake. With the 8 foot walls in place, it simply does not look right. Before proceeding with building the four foot wall to be placed on top, I decide to take the situation home and sleep on it.

By morning it is clear to me what must be done — it is remodeling time. Many new houses undergo significant remodeling while they are under construction and New House will be no exception. The walls must come down while I head out on another expedition in search of 12 foot stud material.

The living room of New House is a single story, 12 feet high. At the back, we need to install a fireplace and chimney chase. We have already framed a space for a pre-manufactured metal fireplace and we decide now is the time to install it. There is only one problem: there isn't sufficient room above the fireplace framing member to permit clearance for the double flue. There was sufficient room until Stan decided a raised hearth would be more in keeping with the feel and tone that he is attempting to create in the living room. So now there is not room enough. Three remedial, remodeling steps are required. Step One requires me to return to the fireplace store for two pieces of offset flue pipe. Step Two requires that a large oval crescent be sliced away from the wooden header to allow passage of the offset piping. Then comes the dreaded Step Three.

Step Three involves putting all the pipe sections together and extending them to a height at least three feet past the highest part of the roof. The problem is that the space into which the piping must go is only open at the top. We have applied all the 1 x 6 rabbeted redwood to the sides, and the pre-fab metal fire box and the framing around it are blocking the bottom.

The first solution we try is to snap all the flue sections together on the roof (Thank heaven it's flat at this area!) and then raise the single twenty foot section of 18" diameter double-walled flue pipe straight up and drop it down onto the offset collars below. That almost works. Four of us do manage to raise the pipe upright and we do manage to drop it down . . . halfway. Then disaster strikes. The weight of the first ten feet of the pipe pulls itself loose from the second ten feet and it crashes down onto the collars below. Needless to say, everything is now badly bent and twisted.

This is bad enough, but now we've got half the pipe down in the hole and half up on the roof. There's no way to snap the second half into the first without either taking the first half out and trying again or else sending somebody down into that long narrow box to put the sections together inside. Carlos is elected. He's the only one of us thin enough to possibly fit.

He's able to climb down to the top of the first ten foot section and successfully put the pieces together. That's a relief — except that now, with the pipe in the way, he can't get out! He can't move up or down and, while he is in no danger, he is stuck in there. Naturally, the first thing we tell him is that it's lunch time and we'll try to figure out something by the time we get back. Somehow this doesn't make him laugh the same way it does the rest of us. But we're laughing out of ignorance. We really don't know how to get him out and keep the flue in.

I should have realized when we were framing the chimney that I needed to leave some portion open for access to install the flue pipe. Actually, to prevent the hassle, I should have had the fireplace on the job and installed it before the plywood and siding.

Finally it looks like the only thing we can do is either cut out the framing and siding in two sections of the flue box, or else dismantle the platform on which the firebox is fastened and resting. We elect to do the latter and this allows us to move the flue and fireplace off to one side, and permits Carlos to come all the way down and out through the bottom opening. This whole fiasco takes more than five hours and, ultimately, becomes very trying and unfun for all of us.

How much time should I generally allow for oversights and errors? There's got to be some, I suppose. But even if extra time is allowed, it doesn't make errors any less of a hassle.

One of my strengths is the ability to solve problems. But some problems are challenges and some problems are hassles. There's a difference between a problem that's a hassle and one that's a creative challenge.

Problems that other people have caused are mostly creative challenges because I'm not to blame and don't feel guilty about them. Problems that I have caused, whether by error or omission, are the real hassles. I seem to be much more tolerant of other people's mistakes than I am of my own. I wonder what would have to happen for me to treat all mistakes equally as creative challenges? This, I decide, is something to seriously investigate.

As trying as the flue dilemma is, it could have been worse. In most building careers of any length, there are bound to be what Fine Homebuilding magazine classifies as "Great Moments in Building History". With regard to the framing phase, one such moment stands out clearly.

On one of my early Bountiful houses Derek had me working with another carpenter named Mark. Naturally, we dubbed ourselves the Marks Brothers, both for our fine appreciation of a good joke and for the fundamentally happy approach we took to housebuilding. I was Mark^4 (Mark IV) and he was Mark^2 (Mark Squared). In recognition of the difficulty in continuing to achieve building perfection, our private motto was: "We build it right . . . the second time!"

One day we decide to propose a challenge to ourselves. We have a forty foot length of straight exterior wall to frame on the second story of the house we're working on.

"Let's see if we can get it all framed," I offer, "complete with door and window headers, double top plates, and plywood siding, and then raise it up into place and have it completely braced and nailed in before Derek comes around for lunch at one o'clock."

We accept the challenge from each other and set to work at a feverish pace. Feverish paces are great fun when everybody is working together at the same blistering speed By noon we have all the stud and header work nailed together and are ready to install the plywood siding. To get the plywood on and completely nailed off requires us to raise the feverish pace even higher. At 12:57 PM the wall is complete and ready to be raised. Only one problem remains: a 40 foot wall, complete with sheathing installed, is very, very heavy — heavier than two guys can even imagine lifting. We had planned to construct and raise the wall in two 20 foot sections but, in the raging haste of installing plywood siding, we nailed over the joint where our break was planned.

"How about this," I suggest. "We'll both lift one end together and get it as high as we can. We'll have braces ready and one of us will quickly prop the brace under the wall while the other holds it up."

"Fine," says Mark[2], "let's do it."

Well, even 20 feet of a 40 foot section of fully-framed wall is very heavy. We only manage to raise the first 20 feet only up to about a 45 degree angle and jam several unsecured braces under it. It then becomes clear that we will have to raise the next 20 feet all the way upright and then go back and raise up the first 20 feet the rest of the way. Okay. We are fevered. We're cooking. And we can hear lunch and Derek's truck making its way slowly up the winding, wooded road.

I get on one end and Mark[2] gets in the middle. We both heave mightily. The second 20 foot section of wall slowly raises into full upright position. As we heave however, I inadvertently kick my braces off the deck onto the ground below. Derek' engine can be heard coming closer and closer.

"I need help here in the middle," Mark[2] calls as I look around frantically for a replacement brace. "I need to get this thing braced and right now." His face is straining a bright red.

Without giving my missing braces any further thought, I let go of my wall, race down past the middle where Mark[2] is lifting and give the first section of wall a mighty shove into the upright position.

For a brief moment the whole wall is raised perfectly into the upright position. And then, just as Derek's truck rolls to a stop, he witnesses 40 feet of perfectly framed wall as it comes sailing, ever so slowly, out over its apex, bends over the edge of the first floor deck and then gracefully crashes down onto the ground fourteen feet below. Without so much as a second look, Derek steps out of his truck, looks at his watch and says: "Sorry I'm late with lunch, guys."

He made sure we built and raised it right the second time.

Thinking about how we had to disassemble, cut up and discard much of the plywood and 2 x 4's from that wall, as well as rebuild it a second time, allows me to put the flue problems at New House in a much lighter perspective.

7

"Pleasure will teach him not to fear
the destructive act;
half the house must be pulled down."
— Constantine Peter Cavafy

Taking It Higher

The first thing I do once the second story walls are braced and in place is nail in a two foot length of simulated rafter and take an exact measurement from the top of the slab to the very top of what will be the finished roof surface. I am motivated by one overiding factor — fear that New House will not conform to the Daylight Plane Ordinance.

During the course of framing the second floor walls, Inspector Dan showed up on the job to pay us an unofficial, off-the-record visit. It seems that in each of the last three weeks Brenda has gone to City Hall several days in succession to complain about New House. Dan relates that she and her architect have trespassed repeatedly onto our property to sneak over and measure all of our slab dimensions, our setback dimensions, the positions of our mudsills, and anything else they can use for information to make a case against us. Dan speculates that the only problem these intrepid intruders could come up with is that our slab elevation might make the house intrude into the daylight plane under the recent changes in the Ordinance.

The recent changes in the Ordinance? What recent changes? We discover that a new ordinance has been put into effect in between the twelve weeks when our plans were finished and submitted for permit, and when our permit was finally issued. The new ordinance could lower the permissable height of New House by a full foot!

The spirit of the DPO is to keep three and four story buildings from blacking out single story Eichler-type and bungalow homes which heavily populate sections of Palo Alto. Normally, with a house the same size as those on either side of it (as with New House), the Building Department is only minimally concerned with incursions into the daylight plane. But because of Brenda's repeated visits and complaints to the department chief, Dan wanted to warn me to make sure that our plane is perfect.

I feel I'm in a double-bind with Brenda. At Christmas she brought over a bottle of champagne for the crew. She also agreed to split the cost of a new fence, and at the very moment of Dan's visit she is selling us electricity from her meter. Then she is clandestinely sneaking over to our property and surreptitiously complaining to the Building Department behind our back. It is clear to me that Brenda and I are in a game of NIGYSOB* and that I am in the ring with a hardball player.

I am determined that Brenda is not going to catch me. I double check our wall heights by sampling all along the length of the top plate. This discloses an average measurement of 18'-3" . . . three inches too high. When I look across the wall over to Brenda's house, I can make out her faint silhouette standing at the back of her kitchen, watching every measurement I make. My immediate paranoid fantasy is that she's standing there with a telephoto lens, snapping large black and white glossies. And now she has the evidence that I'm three inches over the permitted maximum. I am about to be NIGYSOBed. That's the good news.

The bad news is that I'm three inches too high under the *old* ordinance. With the new ordinance, instead of three inches, I am more like 15" too high! This is a major concern because it is much more than we can take off the height of the second story walls. I have already spent many hours and several nights trying to come up with a

* *NIGYSOB stands for — Now I've Got You, You SOB. It's a commonly played psychological game identified by Eric Berne in his classic book,* **Games People Play**. *The internal critical parent and the adapted child in each of the people involved participates, essentially delivering these messages to each other: Parent — "I've been watching you, hoping you'd make a slip." Child — "You caught me this time." Parent — "Yes, and now I'm going to let you feel the full force of my fury."*[1]

creative solution. Finally, the *only* solution I can think of (in terms of proper lumber-mechanics) is to tear down two floors of framing, demolish and excavate the entire slab, lower the grade, and start . . . all . . . over . . . again . . . from . . . scratch!

After work I take the problem home once again. I'm very worried. I have no reasonable solution. I don't sleep well that night or the next, and I'm terrified to tell anyone about the problem, lest it become spread all over town. I'm terrified that people will yell at me and tell me in a variety of ways that I'm stupid and incompetent — that I'm still just a loser kid from the housing projects. These thoughts do not allow sleep in any form. Something has to give. I either solve the problem or we tear down New House's framing and start over.

In the moment my mind feels like it's on fire. I need to quiet these painful thoughts. Over the years I have experienced many events that send my mind into freewheeling dread. The words are different each time, of course, and if I subtract out whatever the words may be, the sound is like a circular saw ripping through plywood over and over again. I have to break the repetitive sound cycle the words make. The only way I have ever found to do that is to first remove my mind and body from the actual troublesome environment. I then seek solace in psycho/spiritual writings. Joseph Goldstein, is one of my favorite authors:

> "The next step on the path is right thought. This means thoughts free of sense desire, free of ill will, free of cruelty. As long as the mind is attached to sense desire, it will seek after external objects, exter-

nal fulfillments which, because of their impermanent nature, cannot be finally satisfying. There is a momentary experience of pleasure and then craving returns for more. The endless cycle of desire for sense pleasures keeps the mind in turbulence and confusion. Freeing thought from sense desires does not mean supressing them and pretending they are not there. If a desire is suppressed, it will usually manifest in some other way. Equally unskillful is identifying with each desire as it arises and compulsively acting on it. Right thought means becoming aware of sense desires and letting them go. The more we let go, the lighter the mind becomes. Then there is no disturbance, no tension, and we begin to free ourselves from our storehouse of conditioning, from our bondage to sense desires.

Freedom from ill will means freedom from anger. Anger is a burning in the mind, and when expressed causes great suffering to others as well. It is helpful to be able to recognize anger and to let it go. Then the mind becomes light and easy, expressing its natural lovingkindness.

Thoughts free of cruelty mean thoughts of compassion, feeling for the suffering of others and wanting to alleviate it. We should develop thoughts which are completely free of cruelty towards any living thing."[2]

The problem is that I don't feel light and easy. I have sense desires. I also feel a lot of anger and cruelty towards, Brenda, Homer and everyone else whom I perceive as thwarting everything that I would like to accomplish.

Only to the extent that I'm able to observe and not act out this anger do I feel the least bit skillful. Only to the extent that I don't behave in ways I have in the past that have produced negative outcomes, do I feel like I'm growing. I feel like I might even be moving forward on the path of learning and Right Thinking that Joseph Goldstein describes.

I decide to do some reading that specifically relates to problem-solving. Russell Ackoff, in his book, *Management in Small Doses*, suggests that there are essentially four ways to handle problems:

- absolution — ignore the problem and hope that it goes away;
- dissolution — eliminate the problem by redesigning the system;
- solution — do something to yield the *best* possible outcome;
- resolution — do something that yields an outcome that satisfies.[3]

I decide to go for resolution. I call the city for an inspection and specifically ask them to come out and check setbacks and daylight plane measurements. I'm worried about the outcome, but I'm willing to have it come out however it will in order to have some resolution. The problem with the daylight plane is potentially a very costly mistake, but still only a mistake, I can't be killed and my daughter won't be forced to wear a Scarlet T-Shirt around town with the letters "LPKO" (Loser-Project-Kid-Offspring) on the front.

Inspector Bob shows up the next day. Like Inspector Phillipe before him, he confirms that our setbacks are on the money. He also informs me the department has determined that because of when our plans were submitted, legally we fall under the jurisdiction of the old DPO. It takes all of Ramon's, Carlos' and Mitch's presence to keep me from kissing him.

We are still three inches too high, even under the old DPO. When I point this out to Bob, he tells me he's been reviewing the plans and is not surprised. He tells me he knows of other houses that have had the same problem. Some have simply cut the walls down. A solution he recommends is for the main sloped roof to have a 2 x 4 cut at a 45 degree angle on the wall and pull the rafters in (*see Illustration 7.1*).

On the flat roof the solution Bob proposes is one that I came up with also: cut the ends of the 2 x 12 ceiling joists at 45 degrees down to the required size (*see Illustration 7.2*).

After Inspector Bob leaves I feel light as a feather. A huge weight has been lifted from my mind. The modifications to the framing will only take me and the crew a day or so to accomplish and we will have resolved the problem of intruding into the daylight plane those three inches without having to go for a zoning variance and without having to tear New House down and begin again. What a relief. And the fact that both Inspector Bob and I have come up with the same solution at the same time calls to my lighter mind the cartoon on page 84.

Much energy and many forces go into building a house. When ideas and events interweave meaningfully, as they

Illustration 7.1

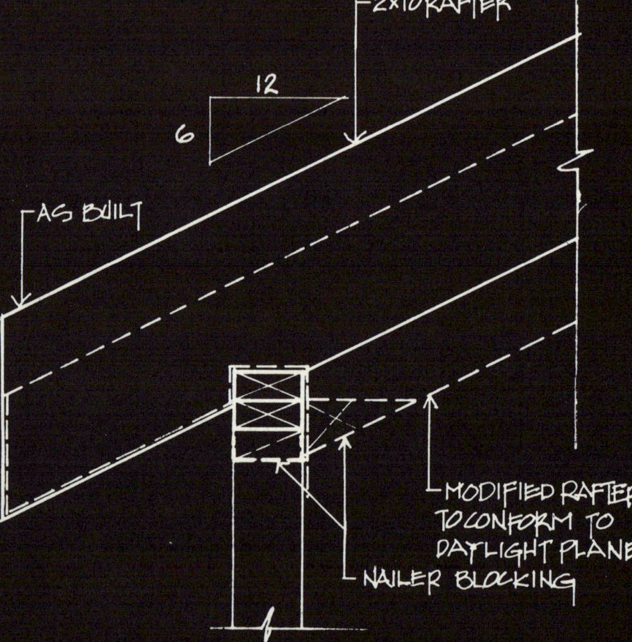

2×10 RAFTER
12
6
AS BUILT
MODIFIED RAFTER
TO CONFORM TO
DAYLIGHT PLANE.
NAILER BLOCKING
EAVE DETAIL
1½" = 1'-0"

Illustration 7.2

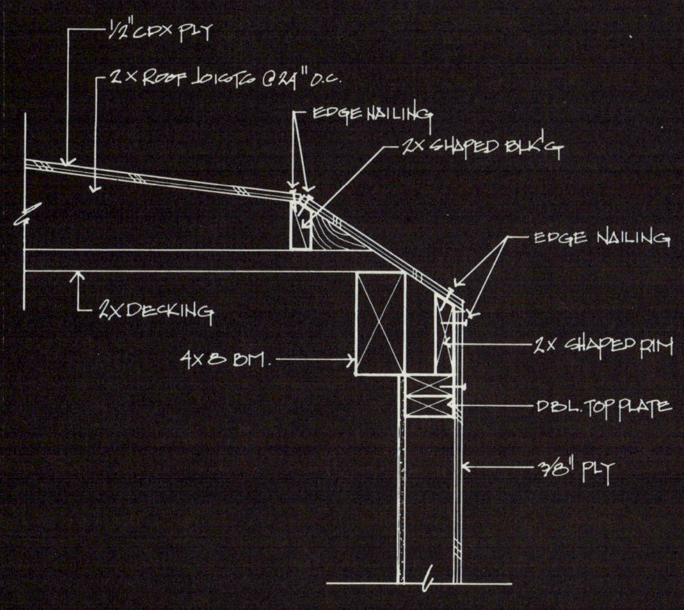

1/2" CDX PLY
2 x Roof Joists @ 24" O.C.
EDGE NAILING
2X SHAPED BLK'G
EDGE NAILING
2X DECKING
4X8 BM.
2X SHAPED RIM
DBL. TOP PLATE
3/8" PLY
A   SHEAR TRANSFER DETAIL
1 1/2" = 1'-0"

*"It's amazing how often two people,
working independently, come up with
the same idea at the same time."*

often do in my life, I am always reminded of the notion
that we are each participants in what Carl Jung called
"orderedness without cause", or synchronicity. It is both
exhilarating and mystifying when synchronistic experiences occur. Here's another example to explain:

Buried in the ceiling of the second floor is a large
wooden beam measuring 15" x 5 ½" x 18'. It's a specially
manufactured beam called a glulam, made of 2 x 6s lami-

nated and glued together under pressure. This beam spans the length of the hall upstairs and carries the 2 x 10 roof rafters on one side, and the 2 x 10 flat ceiling joists on the other side. I am almost ready to order it from Elmer, my main lumber man, when one night I have a dream.

On many of my trips to Palo Alto's main library I have passed by a building lot with a chain link fence around it. Over the years the only work on it has been to install a concrete foundation. Each time I pass I consider checking to see who owns that lot and why no progress has been made.

In my dream I stop off at that lot and discover there is a glulam beam sitting on the slab that is exactly the same size as the one I need. I walk through an opening in the fence to get a closer look at it and try to lift it up. Just as I am thinking about stealing it, a man appears, tells me he is the owner of the lot, and asks if he can help me. He seems friendly enough, but I tell him I was just trying to see how heavy the big beam was. I then quickly exit and continue on my way to the library.

The Saturday after this dream I am driving by the chain-linked building site and I notice some men working there. I slow down and realize one of them is Inspector Dan! Only he's not inspecting, he's doing real work, hammering and nailing. I step in and pretend I am from the building department. I ask to see his permit. He jokingly tells me he has none and then relates a bit of his personal history with this house. He has taken a sabbatical from the building department to try to get the roof on the house before the winter rains. It is then that I notice he does indeed have a large glulam beam sitting on the site!

"Where does that beam go?" I ask.

"It was going to go over the garage here. But it's too small and too short. I bought it as a leftover from Ajax Construction. I thought it would work here, but it turns out it won't. Do you know anybody who has a use for it?"

"What size is it?" I ask, amazed at this close correspondence with my dream. "The house we're doing has a big glulam in it."

"It's 15" by 5 ½" by about 24 feet."

"I'll take it. How much do you want for it?"

"Just pay me whatever you think is fair. Like I say, I bought it leftover from someone else's job. I'd just like to see it used somewhere."

Even though this situation has the magic of synchronicity working for it, the more I think about it, the more I feel like I'm in a bind. I can buy this glulam from Dan, or I can buy it from Elmer, at the lumber yard. Dan wants me to pay him only "whatever I think is fair", while neglecting to tell me what he thinks it's worth. Dan knows New House needs this beam. If I don't buy it from him, how will my inspections go in the future? If I do buy it, there may be a conflict of interest: I'm giving him money and he's doing my inspections. The fact that he's giving me the beam for the money seems somehow peripheral. How can I do the right thing and make life easy for everyone involved.

I mull it over and then one day, the solution synchronistically appears. I am visiting with friends who have other friends over visiting with them. One of the visitors turns out to be the Palo Alto Assistant City Attorney! When I'm introduced, the first thing she says is, "Aren't you the contractor who's buying some lumber from Dan, our building inspector?"

What I thought was a private transaction between two scurrilous criminals, is instead something all the city government knows about! Obviously Dan had done the proper thing and checked out the propriety of the transaction ahead of time. That left no doubt as to the correct course of action. I buy the beam from Dan and put it into our second floor ceiling.

With the daylight plane problems resolved and the glulam beam obtained and installed we can now happily continue with the framing of New House.

8

> "Compassion simply stated is leaving
> other people alone."
> — Ram Dass

Fleshing Out the Frame

New House is completely wrapped in plywood from the mudsill at the concrete slab, to the joists and rafters at the roof. In addition three interior walls have three-eighth's inch plywood installed and nailed very closely to act as shear walls.

Shear walls are a late addition to residential construction in California. The idea is to rigidify key walls with large openings in the house so they will resist the upward and lateral motion generated by earth movement in an earthquake. Nailed plywood controls the lateral movement, and vertical strength is achieved by using anchor bolts to hold the house fast to the foundation. In addition heavy metal straps are embedded in the slab as it is being poured, and these are later nailed firmly to the framing with special heavy gauge nails. Other straps are used to connect the first floor framing to the second floor. Finally, metal twist straps are added to connect the roof rafters to the second floor framing. Supposedly it has been engineered to withstand the greatest loads produced by any of California's daily earthquakes. Perhaps. My question is: what happens when a fissure opens up right through the middle of the dining room?

The plywood that now wraps New House is a marvelous addition to the world of modern building materials. Out of curiosity I looked into the history of its development and discovered that wooden veneered laminate was used more than 3500 years ago! Plywood was used in making furniture in the early 1700's, but it was for use in manufacturing doors in the 1890's that plywood first began being produced in large quantities. The demand became so great that in 1904 the Paine Lumber Company in Oshkosh, Wisconsin constructed a separate facility devoted solely to the mass production of plywood for doors.

Seeing plywood manufactured in a lumber mill is an extraordinary sight. Giant logs of Douglas fir are loaded onto a peeler and, as the log spins, large knives slice eighth inch thick sheets of wood from the log. It's a process very much like peeling the skin from an apple, only on a much, much larger scale. Most impressive is how the layers are cut from the logs with such uniformity in their thickness. Anyone who has ever used plywood knows that the variation in thickness from sheet to sheet is usually virtually indistinguishable.

To install plywood sheathing, many contractors use pneumatic nailers to do the repetitive nailing required. I do not. Nailing off plywood is the perfect pick-up job for either Ramon or Carlos when I am at wits end and cannot clearly plan and direct what other useful work they could be doing. At those times, which are more than I would like to admit, I tell them to go nail off plywood.

With the installation of the plywood, New House is no longer just a naked, skeletal structure. It begins to feel more like a house as it becomes an increasingly enclosed space. We are shielded from prying, critical eyes, and no longer exposed to the curious stares of well-wishers or sidewalk supervisors. There's a place to hide.

There's also a place to hide tools and materials. In the months that we have been working on New House, not a single scrap of material nor a lone, inadvertently left behind tool has managed to walk off the job. Of course we do have a wire mesh fence and a gate across the front (to serve more as a safety barrier rather than to thwart thieves). I recall one near-incident which occurred before that fence was erected.

I am working alone at the site after the lot has been cleared, cutting and hammering in stakes in an attempt to determine the precise layout of New House. I need to buy some extra line to mark the layout so I leave my Skilsaw and electric cord on the ground behind some wood. As I'm driving away a guy in a pickup passes me going the opposite way. He slows down and stops in front of the site. I slow down, turn the corner, and drive around the block back by the site to see what the driver wants. As I pull up to the lot I find the guy loading my saw and cord into the back of his truck!

"Hey, how's it going, " he says, walking over to my truck.

"Not too good," I say. "It looks like somebody is trying to steal my saw and cord."

"Oh, is this yours? I wasn't trying to steal it, man. I was just going to borrow it. I talked to your neighbor there, and I told her how I'm doing this job just down the street and my saw broke and I just needed a saw for about ten minutes to finish my job. I told her I'd bring it right back, so she said that she didn't think you'd mind."

"She didn't think I'd mind? Well, I do mind. Did your cord break, too?"

"Hey, no problem, man," he says, and walks back to his truck. There's menace in his voice. "I'll just put it back where I found it and see if I can borrow one somewhere else. No problem. By the way, you need any help with the framing on this job? I'm a crackerjack framer."

I politely decline the offer of help and when this sticky-fingered, free-wheeling fellow has left, I call my friend Laurie who's a Palo Alto police officer and give her the guy's license plate number. It turn's out the truck is stolen and the guy has a half dozen arrest warrants out on him for burglary and for assault and battery.

This news is somewhat upsetting. In retrospect, I realize he was probably close to assaulting me. There was definitely tension in the air. Had I been the least bit challenging or provocative, that might have set things off. Essentially, this guy was caught red-handed. His story was minimally plausible, but it allowed him to save face even though it was a bungled theft and a bonehead explanation.

Another incident with stolen goods occurred several weeks later when a fellow walks in off the street and asks Carlos if he is interested in buying a brand new circular saw. Carlos directs him to Ramon, as I am off working on another part of the house. He shows the saw to Ramon and asks if he is interested.

"How much?" asks Ramon.

"Seventy-five bucks."

"Too much. I'll give you forty."

"Forty bucks," the guy is screeching. "It's still in the box. These things cost $175 in the store."

All this noise brings me over to see what's going on. By the dirty look of the guy and the shiny look of the saw, it is immediately obvious that the saw and this guy don't belong together.

"Fifty bucks is as high as I'll go," Ramon tells him.

"Wait a minute," I interject. "You got a receipt for that saw, pal."

"Yeah, I got a receipt. I bought it from my uncle."

"Where'd he get it?" I ask, and then I catch myself. "Never mind where he got it. We're not interested. Take a hike."

The guy promptly turns and walks off. I fully expect him to pick up one or two of our tools on his way.

I tell Ramon: "You buy that saw from him and next week someone will be trying to sell some of your tools on some other construction site. What goes around comes around. Karma. Cause and effect. Every act brings certain results. Things don't just happen by accident or by chance."

"But he was going to give me a good deal."

"Not so good for the guy whose saw that used to be. A truly good deal is a good deal for everyone involved."

Two weeks later I am approached in front of New House by a guy in a pickup with a toolbox on the back.

"Look, I'm trying to get some money quickly to help my wife's sister get out here from the east coast. Can I sell you some tools at a fair price?"

This guy is clean-cut and his pregnant wife is in the truck beside him. I notice that his hands are calloused and when I take a look in his tool box, all his tools are used and one of a kind.

"How much do you want for your Sawzall?" I ask. At this point Ramon has sidled up and is looking on with great interest.

"How about $20?" the guy asks.

"How about $50," I say handing him a Ulysses Grant. "And if you need it back, you can have it for the same price."

"Wait a minute. What about karma," says Ramon, giving me a little of my own right back. "How about cause and effect?"

"But I know that guy. He's my brother-in-law," I tell him with a conspiratorial wink.

At the end of the day, I hand Ramon a book by Ram Dass, *The Only Dance There Is*, with the following passages underlined in it.

"I'm working out a way to reduce the price and lower the profit on my book, *Be Here Now*, and I go to my father who is a very wealthy Republican from

Boston with a big estate, and he says to me, 'Well, what are you, against capitalism?'

I said, 'No, I'm not against capitalism. I think it's wonderful. I'm happy you're a capitalist and you are enjoying it, and it's wonderful. I'm happy I do what I do and I don't want to change you. If there are more of you than me, then we have to play by your rules. If there are more of me than you, then you have to play by the other rules. That's the way the game is.'

So he said, 'I don't understand why you are cutting the price when you have a best-seller. Everybody is willing to buy it.'

I said, 'Tell me something — you're a lawyer — you just tried a case for your brother-in-law, didn't you?'

'Yeah.'

I said, 'Did you charge him a big fee?'

'Of course not.'

'What did you charge him?'

'I charged him my expenses.'

'Why didn't you charge him a big fee?'

He says, 'He's my brother-in-law.'

I say, 'That's my predicament.

Everybody is my brother-in-law — what am I going to do?'"[2]

9

> "Forgiveness is acquired . . . Forgiveness is
> the key to happiness."
> — A Course in Miracles

Raising up the Roof Beams

There is a tradition in housebuilding: as the first roof rafters go in a live tree branch is nailed to the top of the ridge. Builders perform this simple ritual as an act of grateful acknowledgement to all the living materials, like the branch, that go into a house. With proper acknowledgement, the work will go well and the workers and owners will prosper.

It is probably a self-fulfilling truth: people who follow this practice and consider it worthwhile, do so with a prosperous mind set. They expect success. They are used to it, know what success looks and feels like, and usually attain it. All of us involved with New House are expecting it will be a success. Just for insurance though, I make sure to hang the lucky horseshoe up over the front door.

A number of incidents over the years have convinced me that it's a good idea to put a live branch on old houses too, particularly those having new roofs installed. A couple of adventures will illustrate:

In the early summer of '78 I find myself ready for some new work. One day a lawyer calls and asks me to give a bid to reroof his house. His "house" turns out to be a three story, 18 room mansion with a 24 in 12 full pitch roof (67° slope!). The roof has four different levels, more than a dozen dormers, and everything is covered with cedar shingles, five inches showing to the weather. I really don't want this job, but I *do* want some work. My immediate gut feeling is to tell this lawyer that I am not interested, that I will be happy to suggest several other roofing contractors. Instead I give him what I think is an outlandishly high bid, and to my unhappy surprise, he accepts it on the spot.

Immediately I figure I'm in trouble for two very simple reasons: First, he accepts my offer right away without any begging or whining. Second, he's a lawyer. I've only done work for four lawyers in my career, and every one of the jobs has turned out poorly for one reason or another. But a deal's a deal, and I agree to do the job. Right away we run into trouble: he want's all the old cedar shingles torn off first.

"But it's not necessary to tear them off," I tell him. "Code doesn't require it, and besides, I have not included tearing them off in my bid."

"I know," he says, "but I checked with my father-in-law, and he says I should have the old ones torn off before the new ones are put on."

"I assume your father-in-law has seen this roof."

"Yes, he has," he replies, "and he says it will make a better job if the old shingles are torn off."

"Well then it's clear that your father-in-law is not a roofer," I say in disgust.

"No he's not. He's a lawyer. He heads the firm where I work."

End of argument. I hire four high school kids to help tear the old roof off, and no sooner is the job started than the complaints begin. Some from the lawyer, but many more from his wife.

"There's too much talking and shouting up on that roof."

"I can't stand all this dirt and dust on the sidewalk, and all over the lawn and shrubs."

"Your men are dropping great gobs of dirt and debris into our attic."

And on and on it goes. Toward the end of the tearoff, one of the kids steps on a piece of 1 x 4 skip sheathing with a large knot in it. The board breaks and sends his foot through the ceiling of the master bedroom. Even though I advised the lawyer and his wife beforehand that with a tearoff of the old roof something like this was bound to occur, the woman goes berserk. It's 11AM and she's still in bed when it happens. She starts screaming that the workmen are trying to spy on her, that we are all perverts, and that she's going to call the cops and have her father sue us for everything we own.

Well, that just about does it for me. I meet with the husband and tell him that my own lawyer has assured me that he will find some way to get me out of the contract should I decide to do so, and that I have decided to do so. Instead of threatening me or using any of the other lawyer manipulations I am expecting, he takes out his checkbook and writes me a check for $500.

"That's for you if you'll stay and finish," he tells me. "And if you can just be patient, I'll take my wife away on vacation for a month. We'll leave in a week."

Every cell in my stomach and brain is screaming for me to give him his check back. But my hand won't do it. I need the money. Instead I put it in my wallet and show up for work once again the next morning.

In this particular house the master bedroom sits directly over the kitchen which it turns out was recently remodeled into a big, fancy, modern country kitchen, with island cabinets, large central breakfast eating area, and all the latest built-in appliances. A voice from a place I call my Spite-Mind assures me that the wife never goes into this room.

I am working on the roof directly above the master bedroom which is over the kitchen. I'm working on a 12" clay flue pipe that protrudes through a custom-made metal, roof saddle flashing. I want to preserve and reuse this flashing, but in order to take it off I need to cut a piece of very rusty plumber's strapping that bolts around the clay pipe and keeps me from being able to lift the flashing off. My tin snips will make short work of that. But, no sooner do I cut the metal strapping, than "woosh" the clay flue pipe immediately disappears from view.

Five seconds later I hear a sickening, deafening crash and I know instantly what has happened: in the course of the previous kitchen remodeling the contractors took out a number of the old clay pipe sections and, since the piping was no longer being used, the contractors simply covered up the hole in the ceiling where they had come through. But they only took out those sections that were easy to remove. Those they couldn't take out they left inside the wall. More precisely, they left them dangling from the saddle flashing, suspended there by the rusty metal strapping — a time bomb waiting to go off.

"They're trying to kill me! They're trying to kill me!"

When I hear the drum-piercing screams from down below, I know immediately this is actually a very ingenious plot the former contractors have cleverly, deliberately and painstakingly put into place as a means of retribution against these people.

Down in the kitchen however, it does indeed look like someone is trying to kill her. Eight feet of hanging, center-aisle ceiling cabinets are torn from the ceiling and are resting in shattered pieces, half on the floor and half on the counter. The gas cooktop is completely twisted, and bent and broken pieces of three sections of dirty, clay flue pipe are scattered all over the kitchen floor. Dust and debris are everywhere. The kitchen actually looks like it has been bombed! The scene is so incongruous, I don't know whether to laugh or cry. Her hysterics make it seem funny, but the thought of having to clean up and restore everything to order is depressing.

Fortunately, the next day the lawyer and his wife go away for a month. When they return my work on the roof is finished. As some small luck would have it, the lawyer and I use the same insurance broker for our insurance and, when I tell him I want no part of the kitchen repair, the broker arranges for some other poor soul to go over and repair the damage. But not, unfortunately, by the time that dynamic duo return home from vacation. My heart goes out to that remodeling contractor. I hope he's hung a live tree branch or a horseshoe over the door.

Another incident that convinces me I should mark older houses with a branch takes place in the fall of 1983.

The San Francisco Peninsula area (where I am doing most of my remodeling and construction) is considered by many to be one of the major psychotherapy centers in the world. Not surprisingly, I do a lot of work for psychiatrists, psychologists and social workers. I have just completed a bathroom remodeling job for Dr. Pete, a local psychologist, and as I am walking out the door he asks if I would build a new gate for his driveway. I decline politely, telling him that I have commitments for the next three months, with several more months work pending. These are prosperous times.

"You need to hire some help," Dr. Pete suggests.

"I would, but good help is so hard to find. Not only that, but I'm pretty finicky about the quality of work that goes out under my name," I tell him.

"That's no problem. I'll make you a deal," Dr. Pete offers. "You advertise for some help and I'll interview and give psychological tests to the people who apply. That way I'll make sure you end up hiring someone we're sure can do the job for you. In return all you have to do is fit my gate into your schedule."

In the next week, a small ad in the local paper for a carpenter's helper produces close to 200 responses! This is before Ramon, and since English is the only language in which I am even modestly fluent, I am forced to immediately cut the list of applicants by 75%.

"Send the remaining 50 a copy of this Skills Inventory," advises Dr. Pete. "From those that send it back we'll find you the perfect helper."

Only two send the inventory back.

Of those two, Dr. Pete sings the highest praises of Dan.

"Dan is your man," the good Doctor assures me. "He's smart. He's athletic. He' s had experience. The worst that will happen is that he may lose a few of your tools. When can you start on my gate?"

Well, it turns out that Dan actually *is* a good man. We finish Dr. Pete's gate, do several small additions, and even reshingle a few roofs together. A small assortment of hand tools — tapes, pliers, screwdrivers, and other tools of that sort — do seem to occasionally find their way into Lost-and-Never-Found-Land but, on the whole, I am quite pleased with Dan's work.

I have been meaning to tell this to Dr. Pete when, one day, I decide to send Dan out on his own to start a new roofing job while I finish up the one we're on.

"Here's the address of the house we're going to do next," I tell him. "I want you to go over there and begin tearing off the old cedar shakes that are there now. Do you know where this house is?"

Dan assures me that he does and that there is no problem. As he drives off I remark to myself that of course

Dan would know where the house is: it's right across the street from Dr. Pete's gate that he helped me with on his first job.

At precisely noon, Dan returns all full of smiles. I come down off the roof to join him for lunch and he proudly announces:

"You're not going to believe this, but I've got almost the whole back of that roof torn off. I'm not going to be able to work much longer, though. Those *black asphalt shingles* are just too hot and they all stick together..."

That evening I ring the good Doctor after dinner.

"Dr. Pete, remember that guy you tested and told me to hire?"

"Yep"

"And remember how you assured me that your Skills Inventory was as precise a measuring instrument as my folding rule?"

"Yep."

"And remember how you told me that the worst I could expect would be for him to perhaps lose a few of my tools?"

"Yep. I remember."

"Well, Pete. Dan has lost a few of my tools, its true. He's also been a pretty good helper. So good in fact, that I sent him out alone to start the tear-off on my next roofing job."

"And ..." Dr. Pete can feel the other dropping shoe on its way.

"And ... he ... tore ... the ... shingles ... off ... the ... wrong ... roof, Pete!"

For some reason this strikes Dr. Pete as uncontrollably

funny. After about five guffaw-filled minutes he is finally able to come back on the line.

"Well, what are you going to do?" he squeezes out between intermitent chuckles.

"I'm not sure," I reply. "We both agree that people have to be responsible for their mistakes. Builds character, cleans karma, et cetera, et cetera. Right?"

"Right." Dr. Pete affirms, chuckling just a little bit less.

"Well then, the question we should ask, Pete, is not what am *I* going to do, but what are *you* going to do: the roof Dan tore the shingles off of is on top of your house!"[1]

As I recall them now, these incidents offered special teachings for me. When I was stuck in the middle of those jobs, all the different aspects and every confrontation was very serious and took a lot of my time and energy. I took them home with me every night. There was little I could do to get away from them. Later, when the work was done, I was able to see them more from a distance. They weren't so serious. I could laugh at them, even at the lawyer and his wife. I could also feel for the people. I know how badly I feel when I'm just a little upset. I imagine how it must feel to be so greatly upset all the time. Having them removed as far away memories, I can now relax and feel forgiving. Even with Dr. Pete, who became very upset when I pointed out that a responsible action would be to pay part of the cost to have his roof replaced, I could be forgiving when the incident was over and done with.

But in the midst of crises, it's very difficult to stand back and sustain feelings of forgiveness and understanding. More than difficult, it seems darn near impossible.

Often, I feel like I'm under attack and must defend myself. To make growth-oriented changes requires a lot of practice. Luckily I know I'll manage to construct abundant opportunities for practice in the months and years ahead.

Together with live tree branches and an occasional horseshoe that I now place on all work, new or old, I frequently take up something I call the practice of defenselessness. Much that I have read suggests there is great safety in such an attitude when I can cultivate it and give it proper exhibition. I need no longer feel threatened by my perceptions of a changing world — a world which provides no safety, that I can't control. One where no peace of mind seems possible, where danger seems to frequently threaten.

When I feel threatened or under attack, by the lawyer's wife, for example, I feel both angry and weak at the same time. My mind is confused, and I rarely know where to turn to find escape from its horrid imaginings. On occasion, I feel totally out of control, like I am possessed by a madness in a form so grim that hope of sanity seems beyond possibility. The sense of threat is deep and far beyond the frenzy and intensity I can imagine in saner, more peaceful times.

But the thing I am most afraid of and defensive about in my mind, is always so much less threatening in real life, by the light of day, out on the job. The real world permits defenselessness. It's not necessary to lie, hide or suppress who I am and what I have to say. Realizing this, I begin to recognize that in true defenselessness is strength. I glimpse an energy and strength in me that is acceptable to express. It's connected to the part of me that loves using living tree branches and lucky horseshoes to acknowledge myself and all other living things. It's the part of me for whom raising up the roofbeams is a simple act of joyous creation.

10

"By wisdom a house is built, and by
understanding it is established."
— Proverbs 24:3

Building Warriors

I am convinced that all the tradesmen working on New House are provided the lessons necessary for achieving sainthood. A famous teacher of Tibetan Buddhism, Chogyam Trungpa, has described such tradesmen as warriors. The qualities that he ascribes to the warrior, for example, could easily describe several plumbers I have known.

First, a warrior must work to create an enlightened society. That certainly seems to be plumber's work, both literally and figuratively. When I have a plumbing emergency and the plumber comes and frees my stopped toilet I become a pound or two lighter soon after he leaves.

Then, without that gastrointestinal pressure working on me, my mental state lightens up even more.

Next, a warrior must work to overcome arrogance. No work necessary here. The very nature of plumber's work carries a built-in anti-arrogance factor. How arrogant can you be in polite company when you spend the largest part of your waking life grunging around every day with the technology of human waste?

Next a warrior must demonstrate the ability to acknowledge fear and move beyond it to fearlessness. I've seen this happen time and again. All it takes is for a gas or water line to break unexpectedly. After the initial fear of flooding the place and/or blowing it up is acknowledged, it's immediately off in search of a bucket and the shutoff valves. Then it's into action to effect a quick repair.

A warrior must also seek to discover the basic goodness of all beings. This could be the stopper for many potential plumbing warriors. It requires that he recognize all people are basically good, or at least they were at one time, perhaps even so long ago as birth. Somewhere along the way events may conspire to cover up much of that goodness. Basic goodness is not one of the qualities that people tend to manifest in the midst of a plumbing emergency.

Finally, the warrior's duty is to generate warmth and compassion for others. Again, the nature of plumber's work and training serves to place him frequently within this realm. The following story will help to illustrate:

On many projects I undertake, I try to do as much of the work as I can. On a small fill-in job one day, I am doing some minor remodeling work for a very sweet lady-friend. The job involves a small bit of plumbing work and

I am delighted to be able to practice my training as a warrior. I perform the necessary repair of replacing two faucet shutoffs with only a minimum of frustration.

That evening as I am having dinner the phone rings. My friend is on the line. She explains that just as she was about to go out, she heard a pop and a hiss and looked under her sink to find water spraying all over the place. She managed to find the main shutoff but would like me to come and repair the problem as soon as possible.

I immediately return to her house and find that a compression nut on one of the shutoffs was apparently machined out of a defective piece of brass and has split completely along one side. Or I may have tightened it too tightly. As I go and purchase a replacement I consider how rare such occurences must be, and how fortunate that it chose to fail seconds before my friend left the house. Ten seconds later and she would have come home to a house filled with water and I would have been on the phone to my insurance man, with any semblance of warriorness vanished as I try to negotiate a reasonable settlement. At any rate I easily replace the compression nut and return to my dinner.

Next morning my friend is on the phone again: her kitchen drain is stopped up. Even though I have done no work on that drain or any other, the failure of the plumbing work I did the day before is very fresh in my mind. I immediately begin reviewing what I could have possibly done to create such a condition. The best I can come up with is perhaps I inadvertently dropped a tool, or a rag, or my imaginary black iron warrior's belt, down the drain.

When I get to her house I discover that not only is her

kitchen drain clogged, but every drain in the house is backed up. Clearly this is a job out of my minor league. This requires a full fifth-degree pipe-wrenched, plumbing warrior. I get on the phone and call Jorge. I'd rather let Jorge do it.

Jorge arrives and immediately determines that the main sewer pipe leading out of the house is clogged. Since there is no cleanout in that area, Jorge suggests that it be dug out, the pipe cleared of the blockage, and a sweep and cleanout be installed. When warriors speak, trainees listen.

He starts digging and in a couple of hours has located the main sewer pipe. Using a snap cutter, he opens up the pipe and immediately the hole fills up with the most vile, stinking water I have ever smelled. It doesn't faze Jorge, in all his warriorness, one iota. He takes out a small electric pump, jumps down into the sewage-filled hole and simply begins pumping it out.

When the water is low enough for him to run his snake into the pipe he determines that the blockage is about 30 more feet down the line. The problem is that he is unable to dislodge it with his rooter. So he moves down the line thirty more feet and repeats the process all over again, sewage filled hole and all. At last he's able to locate the blockage. With my friend, several neighbors and myself standing over him, Jorge at last comes up smiling. In his two hands he's holding the blockage, which my ladyfriend sees and turns the brightest, beetest red I've ever encountered: it's a 4″ diameter by 2′ long, root-encrusted bundle of used condoms!

Though all the people who work in the trades that go

into the making of New House are offered the training necessary to become Warriors, plumber's work seems toughest. To be a plumber means you have to be willing to crawl around in the dirt under people's houses, it means you have to go into their bathrooms and literally deal with their piss and crap. It means you have to work with dirty, smelly tools and materials. Finally, you have to be able maintain an extraordinary high tolerance for frustration — a factor which Tony Robbins (of Optimum Performance Technologies®) considers Key Number One to everybody's door to wealth and happiness:

> "If you want to become all you can become, do all you can do, hear all you can hear, see all you can see, *smell all you can smell* (italics my addition), you've got to learn how to handle frustration. Frustration can kill dreams . . . Frustration can change a positive attitude into a negative one, an empowering state into a crippling one. The worst thing a negative attitude does is wipe out self-discipline. And when that self-discipline is gone, the results you desire are gone.

> "So to insure long-term success, you must learn how to discipline your frustration . . . The key to success is massive frustration. Look at almost any great success, and you'll find there's been massive frustration along the way. Anybody who tells you otherwise doesn't know anything about achieving. There are two kinds of people — those who've handled frustration and those who wish they had."[1]

The above description aptly portrays Jorge. He's financially secure, he's happy and he has clearly handled massive amounts of frustration in his work and in his life. For more than a quarter of a century he's run a successful, service-oriented plumbing business. He's a good man, a good plumber, and does good business. Jorge probably wouldn't describe his work as a "conscious business", but the fourteen lessons Mirabai Bush formulated in running her business are clearly reflected in Jorge's:

1. See your business as an organism, alive, always changing — a process, not a static institution, requiring continual and spontaneous awareness to be managed effectively;

2. Practice compassion and empathy;

3. Keep close to the ground. Use common sense and simple solutions;

4. Work on creating a happy workplace;

5. Care about what you produce. Make sure it is in harmony with your values;

6. Continually rearticulate the values of the business;

7. Create effective strategies — it helps to remember that business is a game. It also gives flexibility and strength;

8. Always tell the truth;

9. Don't get in it just for the money;

10. Keep the rest of your life alive and diverse. Remember your other priorities;

11. Encourage personal growth. It increases productivity;

12. Hire people you trust, people who share your values;

13. Know that the ends never justify the means. The

process is it. The ends are conditioned by the means;
   14. Be playful.[2]

Leon Pyle is not a plumber, but he too, is a warrior. He's also a friend of mine who recently started his own building company. He builds tract houses in Contra Costa County, California, and he is attempting to infuse the warrior philosophy into his whole organization. The following are a number of the guidelines and values which underlie his new company, Sterling Homes:
   • We are in the business of relationships.
   • We respect our customers, listen to their desires and to their complaints. Customers are our teachers. They will teach us how to refine our products and our systems.
   • We seek win/win solutions to every problem.
   • Subcontractors and suppliers are part of our company, and their contribution to our company is respected as much as our own.
   • We are a learning community, teaching each other in a non-threatening environment.
   • We encourage different opinions.
   • Early communication regarding lack of performance is a form of caring.
   • Honest praise is a more effective motivator than honest criticism.
   • Defensiveness has no place in our company.
   • We are a service organization.
   • We will always err on the side of doing too much, rather than too little for our customers.
   • We keep our eyes and ears open for customer dissatisfaction and take proactive steps to increase satisfaction.

- We are strongly biased towards action.
- We allow for mistakes. We celebrate good tries and small failures.
- We prioritize and deal with what is truly urgent at once.
- We pay equal attention to big ideas and little details.
- We express appreciation to subcontractors and suppliers.
- We value enthusiasm and the ability to create a pleasant work environment.
- We do the best we are capable of at all times.
- Integrity is doing what we say we will do.
- The way customers describe us *is* who we are.[3]

Obviously these precepts serve as guidelines, as ideals to which people can aspire. People and institutions that model and encourage these attitudes and aspirations can be found all around us. For example Leon (Sterling Homes' president) went to a college modeled after an alternative elementary school. And Jorge lives across the street from a nursery school:

"The kids are wonderful to be around," says Jorge. "They are great teachers themselves. They love to learn, they love to play and they tell the plain and simple truth about most everything. They trust themselves, the other kids and the teachers. They're not worried about things like time, and money and getting projects over and done with. They just grow and learn and do things for the doing. That's how I try to do plumbing."

Robert Fulghum has popularized Jorge's experience and insight with the following list of his own preschool learnings:

"All I really need to know about how to live and what to do and how to be I learned in kindergarten. Wisdom was not at the top of the graduate-cshool mountain, but there in the sandpile at Sunday School. These are the things I learned:

Share everything. Play fair. Don't hit people. Put things back where you found them. Clean up your own mess. Don't take things that aren't yours. Say you're sorry when you hurt somebody. Wash your hands before you eat. Flush. Warm cookies and cold milk are good for you. Live a balanced life — learn some and think some and draw and paint and sing and dance and play and work every day some.

Take a nap every afternoon. When you go out into the world, watch out for traffic, hold hands, and stick together.

Be aware of wonder. Remember the little seed in the Styrofoam cup: the roots go down and the plant goes up and nobody really knows how or why, but we are all like that.

Goldfish and hamsters and white mice and even the little seed in the Styrofoam cup — they all die. So do we.

And remember the Dick-and-Jane books and the first word you learned — the biggest word of all — LOOK."[4]

This is a fitting description of both Leon and Jorge. In my experience both are warriors rejoicing in working, living and learning at the kindergarten level. It's something to which I can aspire.

11

"When I tire of the treachery of Reason,
God knows I'm grateful to my heart . . ."
— Khalilullah Khalili

The Heart
of the House

New House is served by a 200 Amp Square D breaker panel. This is a larger service than is generally installed in most new houses. Our thinking is that with advances in electronics, computers and electric appliances occurring quickly, the electricity requirements by the year 2000 could be considerable. Whatever the requirements are, the New House of 1987 will be prepared.

Current city ordinances require that electric feeder cable be installed underground. The rationale is that in the not-to-distant future, all the telephone poles in America will eventually be taken down and replaced by underground service. It's doubtful that it will happen in my lifetime, but at least I'm making my contribution.

Wiring a new house is one of the more enjoyable tasks. Originally I had planned to assign this play-work to David, a licensed electrical sub. Even though he has given me a bid and signed a contract (and is thus required by law to perform), he informs me that his schedule is so crowded that it will be a hardship for him to devote the required blocks of time to the job. I think about this for a few days, talk it over with Mitch, Stan and Ramon, and decide we will undertake the wiring work if David will act as a consultant on an hourly basis. This approach will help our budget and I feel confident that I could do a competent job on the wiring, although I'm not an electrician. I'm constantly aware that it's important to remind myself that I don't know what I don't know. Forgetting this can be particularly dangerous and costly with electricity.

David likes my proposal and agrees that, between me, him, Mitch, Ramon and the city electrical inspector, we can probably do a competent job. Ramon, who has considerable experience as an apprentice electrician in Uruguay, will be the principal person in charge of the work. He and I make a friendly side bet for $10: I say that he will not be able to install all the electrical wiring without some problem cropping up. He says he will. I feel confident that I am a shoo-in on this wager. Either way I'm a winner.

Ramon's work proceeds evenly and without incident. I've installed enough small electrical work and built enough houses to be able to tell when something *looks* out of whack. I feel confident the job will be fine. David visits periodically and he too, confirms my observations.

When Ramon's work is complete, we actually manage to pass rough electrical inspection. Later we even manage to pass the final electrical inspection and obtain the "Permission to Connect Service" sticker. The city utility men come and, with a minimum of difficulty, manage to hook up our 200 amp meter main outside the garage. This panel feeds another circuit breaker panel in the utility room. All the breakers there are in the "off" position, and the utility room main breaker switch is off, too. When the house meter main is off the whole electrical system is off. Switching the house main on only sends power to the utility room breaker panel. Sounds good and right.

When I switch on the house main however, I notice that the little aluminum wheel in the meter begins to spin quite noticeably. Not only that, but I can hear the distinct hum of kilowatt hours being rapidly consumed. I immediately think the main switch on the utility breaker panel isn't in the off position after all. Except when I check that switch, I find it is off. I go back and shut the house main off. The hum stops humming and the wheel stops turning. Once again I turn it on and go back to the utility panel and turn all the circuits on and off and flip the main breaker back and forth and make sure I leave it the off position. Still the hum keeps humming.

At this point I can feel negative magical thinking beginning to make creepy inroads. Too many Stephen King novels. Though it is our creation, this house has a life of its own and has now taken over control. When I return to the meter main, I half expect that flipping the switch will no longer stop the wheel from spinning and the hum from humming. But it does, and rationality soon returns . . . in the form of Ramon, fresh from a dental appointment.

"Let's find out where the hum is coming from first," suggests Ramon.

It turns out the hum is coming from the Jacuzzi motor in the whirlpool tub up in the master bath. We pre-tested this motor and had to replace and repair it twice before we could tile up the tub. Now, nothing we do except flipping the meter main will shut this motor off.

"Somehow, the neutral or the wire that we used to bond the motor to the meter main ground as a safety precaution must be coming into contact with a hot wire somewhere," Ramon theorizes. "If we disconnect the neutral wire, the motor should turn off."

So that's what we do.

We disconnect the white neutral wire and . . . the motor keeps humming. Doo doo, doo doo. Doo doo, doo doo. Doo, doo, doo, doo. More magical thinking. I'm thinking about getting out of there.

"Well, if we disconnect all the wires, the motor should turn off," says Mitch.

"What if it doesn't? What then?" I challenge, feeling ten years old.

"Let me think about this for a minute," Ramon intervenes coolly.

Minutes later he is patrolling the house with the old Ohm meter and nodding sagely to himself. A short time passes and he calls us into the utility room.

"I've got it. I know what the problem is," he says. " .

. . This house is haunted." My worst fears are immediately confirmed. Tons of time, money and energy all these past six months, completely for naught. My throat gets very tight.

"Who pulled these three heavy black wires from the meter main, and who wrapped the white tape around the neutral?" he asks.

I look up, down and around deliberately avoiding his gaze. My throat gets a bit tighter.

"Well, whoever that person was," he says, staring directly at me. "he somehow got the hot wire connected to the neutral bar and vice versa. When the panel is off 110 volts are going out to all the plugs and lights through the neutral without going through the breakers. When the system is on, *220 volts* are going to half the plugs and lights in the house! The only reason it wasn't obvious is because the Jacuzzi tub motor is the only appliance that we have hooked up right now. We can't shut the Jacuzzi off because the power is coming unswitched directly through the neutral!"

"Geez," I finally manage to mutter in a half-hearted defense. "Those utility guys sure messed this one up."

"Or," Mitch chimes in, "maybe the joke-playing ghost of some long dead electrician is at the heart of this matter."

With the spooky Jacuzzi behind us, one day at lunch I pose this question: "What would you consider to be the heart of a house?"

"I'd have to say the heart of this house is the kitchen." Mitch answers. "It's the center of all the action, especially with the kitchen, hall, breakfast nook and family room completely open the way it is. This is 'Actvity Central'. Cooking, cleaning, eating, entertainment. It all takes place right here inside this 800 square foot space. We're standing at the heart of the house."

"I see it a little differently," says Ramon. "I look around and I know all my wires are running through this house. They stretch out all over and carry energy from the crawlspace all the way up to the attic. From the underground area out back, all the way to the upstairs master bedroom bath.

"To me the black hot wires are like arteries carrying the energy around to wherever it's needed, and the white neutral wires are veins making the return trip. And at the heart of all this activity I see the electric service panel. Just like a human heart, if the electric panel shuts down, the whole house is out of commisssion."

"And I suppose once the house is shut down," says Mitch, "the green wire buries us in the ground?"

"Correct," Ramon replies. "And when I think about how Tesla and Edison feuded over the merits of alternating current and direct current and how Tesla proved A.C. to be superior, the comparison carries even further. Just like in the human body the current that runs through my electrical system has a continuous and constant pulse, and at the center of that pulse is the service panel. And just the way people experience heart break, my panel has heartbreak controls already built right in. The circuit breakers!"

Needless to say this discussion continues on and extends lunch much longer than necessary. But there are further analogies to be made and lessons to be learned. Several days later I reconvene our discussion.

"The heart of the house for me," I announce, "is the people. All the people that we come in contact with from the initial purchase of the land to the final sale of the completed building. They all bring their energy to bear. They are at the heart of the house."

"You know," says Ramon. "I recently read where some builder estimated that in the course of buying, building and selling a new house he came in contact with nearly 2000 different people.[1] That seems like a lot of people."

"It is. And the relationships that we develop with each of them form the strands of a structure that make for one big association."

"Uh, oh." says Mitch. "Here it comes. Next thing you know, he's going to be hitting us up for a contribution to his heart association!"

"Seriously," I reply. "That's where the real fulfillment comes from — my involvement with all the other people. There is very little that I can accomplish by myself. To be successful I have to involve others. And I have to be just as concerned about their success as I am with my own. If I'm successful with this house and the 2000 people who were involved on its construction with me are not, what kind of success is that?"

"Obviously you're talking about more than financial success," says Mitch.

"Making decent money is important. But equally important is the strength, integrity and the workability of the relationship. Some relationships are brief and cordial. Some are brief and antagonistic. Most of them I can only do my best to give and keep my word about all that I will and won't do. From there the foundation is laid for other possibilities to develop."

"From there," says Ramon, "you can begin to develop strength of heart."

12

"Ah, to build, to build!
That is the noblest of all the arts."
— H.W. Longfellow

Experimental Fridays

From the day I first witness my friend Derek's abrupt transitions from housebuilder to pre-med student and then back again to housebuilder, I am consistently drawn to the various aspects of the craft. His Sufi teacher's directive suggested that housebuilding had the potential to be particularly noble and honest work. The more of it I do, the more I begin to see the potential for housebuilding to provide the "perfect vehicle for the perfection of the human character." I feel sure his teacher saw housebuilding as "good work". Good Work as defined by E. F. Schumacher is performed with three main purposes:

"First, to provide necessary and useful goods and services. Second, to enable every one of us to use and thereby perfect our gifts like good stewards. Third, to do so in service to, and in cooperation with others, so as to liberate ourselves from our inborn egocentricity."[1]

It is from the Buddhist principles of "Right Livelihood" that Schumacher has drawn his notions of "good work". From the very first days of learning and building with Derek and Bountiful Builders, I have continued to look for ways that I might operate by employing such principles. I want to build houses with heart. It is difficult to separate the principle of Right Livelihood from the whole of Buddhist philosophy. But I do it anyway. Thus far here's what I've come up with.

Right Livelihood work is unpretentious. It promotes opportunities for growth and learning by permitting experiences along the whole emotional spectrum: joy, frustration, sadness, fear, excitement, anger, etc. It's also work that contributes constructively to to the well-being of society. It is work that is particularly right for me at this moment in my development. (And the same work may not be right for me at another time, which is important to remember, but difficult to recognize.)

Les Black and Don Stone, a lawyer and an accountant respectively (of all things!) call their practice of Right Livelihood, "Whoopee Work".[2] For them, fun is at the core of Right Livelihood. You have found your place when you awaken with a day of work ahead and say, "Whoopee!"

For much of my life and for most of the people that I know, the days that I awaken and say, "Whoopee", are the days that I don't have to go to work! More often I feel like the guy in the cartoon on the following page.

But I've never closed myself to a different experience as a possibility. Now, as a builder, I'm finding more and more days that feel fresh and interesting, days when I'm genuinely excited and filled with anticipation to get to work.

In order to keep our work on New House fresh and interesting, I encourage people to experiment, to try out new things. To that end we officially designate the last Friday of every month as a day for experimentation. On that Friday we all have permission to experiment with new ways of working and being.

Ramon tries using different tools for jobs other than they were intended, like using a roofing hammer for framing, or a reciprocating saw to do the cutting usually done by a circular saw. Carlos, who is normally right handed, attempts to work the whole day primarily left handed. Mitch uses one "Ex-Friday" to see if he can get through the day without talking. On another Friday, he goes all day without food to see what working with no nourishment is like. One day the whole crew walks around whispering a repetitive chant while they work: "Sorry. Sorry. Thank you. Please."

For several Ex-Fridays I attempt to do everything with extra awareness and deliberation. Ramon calls those Fridays my slow-motion days. His theory is that I have chosen this experiment because Friday is payday and, if I go slower, the day will last longer and I won't have to pay

everybody so soon. Zen Buddhists might call it my day of mindfulness, using a quality of mind that notices what is happening in the present moment with no clinging, aversion or distortion.

My mindful Fridays begin upon awakening. On Thursday night I leave a note in the journal by my bed: Wake-Up Call. The call to wake up starts simply with the directive to slow down. I begin the day by getting out of bed and getting dressed with as slow and deliberate a motion as I can manage. It's a practice that I refer to frequently on the job as "paying attention to the relevant dimension". On Ex-Fridays the volume control for that dimension is turned up considerably.

Charles Tart considers an experimenter like me a "successful and perceptive malcontent", and describes such an orientation in his book, *Waking Up*:

"The successful malcontent is not content with the level of success he has attained in ordinary life . . . he has tasted the rewards society has promised for being normal (respect, consumer goods, security, etc.) and the rewards for rebelling, and has found that while the rewards for either may be pleasant, they are not enough. Something vital is missing . The perceptive malcontent realizes that more material possessions or social respect or the romance of rebellion will not satisfy this need for something vital."[3]

J. Krishnamurti, an Oxford-educated, Indian philosopher, describes the same attitude as "creative discontent":

"Only through real discontent does one have initiative . . . one must be wholly discontented, not complainingly, but with joy, with gaiety, with love . . . and as you grow older and keep your discontent alive with the vitality of joy and great affection, then that flame of discontent will have an extraordinary significance because it will build, it will create, it will bring new things into being."[4]

Experimental Fridays are the beginnings of the malcontent's search for the missing vital. The first unmistakeable thing I notice each Friday that I begin this practice is that I feel better. All sense of urgency is reduced. Moving with slow deliberation induces tranquility and allows me to pay much closer attention to all that I am doing in the present moment. I find my mind wandering off less frequently in search of disastrous futures to guard against.

My daughter, Amanda, notices the differences on these Fridays immediately. She is able to finish her cereal before me and raise her spoon with a triumphant, "I beat". She also notices that I am much more relaxed and don't have much to say on that day. What I am doing is spending a lot of time simply noticing my breath going in and out through my nostrils. That or whatever else happens to be the task at hand.

Fridays tend to be days on the job designed to restore order that has been neglected for the rest of the week. I slowly organize stacks of lumber, sweep or make lists of jobs that need to be accomplished the following week. I tally materials that need to be ordered or purchased. I also spend time detailing costs and attempt to anticipate what expenses will be forthcoming. To the extent I am able, all these tasks are done slowly and evenly without any sense of reluctance. No job is undertaken from the single desire to be done with it, to simply have it be another item checked off the list of things to do. Rather, each is done from the perspective of the desire to be present, aware and awake in each moment as it passes.

How does it feel to spend the whole day being mindful?

First of all: what "whole day"? I probably spend ten percent of any Friday being awake. The rest of the day I'm lost in space as usual. Thinking about yesterday, thinking about tomorrow. Thinking about work. Thinking about this, that and the other thing. When I deliberately attempt to spend significant amounts of time in the present moment I am astonished to discover just how little time I actually spend there. Most of my time on Fridays is spent recognizing that I have drifted away from my experiment in mindfulness and directing myself to return.

And I also discover there are very different experiences of present-moment awareness. The exhilarating awareness present in framing a wall as quickly as possible and the contemplative awareness present in slowly sweeping a pile of sawdust into a dustpan is somehow very different. One experience is speeded up and one is slowed down, and for me slowest is hardest. My mind wants to race and drag my body along with it. Placing a mental governor provides the deepest sense of tranquility and peace. I can relax and bask in the realization that I'm generating my own discomfort — and that it's not a requirement of the work.

With each of these ways of working, I also realize the more I am able to be fully present in a given moment the greater the depth of my experience of being alive. What I seem to perpetually discover is enormously profound in its simplicity: at any given second in time, at any moment or instant, if I take a snapshot or a cross-section and examine my existence in detail, I realize that I am perfectly alright. I am safe.

With awareness, I am safe from all the horrible possibilities that perpetually arise in the imagination: cutting off my fingers, falling off a roof, being yelled at by friends, lovers, therapists and building inspectors, being dunned by bill collectors. I can and do walk through the world unharmed.

From a place of feeling safe I can better carry on the dual process of self-observing and self-remembering. Tart describes what is necessary for both:

"Your ability to observe yourself is a combined function of your desire to observe yourself, opportunities for observation, obstacles to observation, and the availability of special aids to observation . . . In its most general form, the practice of self-observation is simply a matter of paying attention to everything, noticing whatever happens, being open-mindedly curious about all that is going on. This everything will almost always be a mixture of perception of external events and your internal reactions to them. You should drop all a priori beliefs about what you should be interested in, what is important and not important. Whatever is, is an appropriate focus for observation . . . The attention that we should strive to pay to our world and our selves is an emotional attention, and a body attention as well as an intellectual attention."[5]

"Self-remembering is the term for gathering our dissociated faculties into a more unified whole. It involves a deliberate expansion of consciousness such that the whole (ideally) of your being, or at least aspects of that whole, are kept in mind simultaneously with the particulars of consciousness. It is remembering our body and our instincts and our feelings as well as our intellectual knowledge, and so it promotes development and integrated functioning."[6]

Without instruction and practice, the experiences that

Tart describes are admittedly difficult to understand intellectually. They need to be done in actual practice — over and over with a sense of newness and interested inquiry.

Which is ultimately the purpose of Ex-Fridays. Once a month a determination to try something new. Something different. Something strange. Something that stretches us and reminds that there is more to life, and housebuilding, than just building houses.

13

"... and they all lived together
in a little crooked house."
—James Orchard Halliwell

Quality Control

The City of Palo Alto deserves its reputation as being a tough town in which to build. They have a multitude of ordinances that supercede and amend the Uniform Building Code, along with tightly controlled zoning, earthquake safety regulations, energy conservation regulations, an Indian Artifacts Archeological Board, and an Architectural Review Board to regulate aestetic requirements. They also have a building department with nearly 20 inspectors and administrators, and this is a small town with less than 25000 households, including a large portion of Stanford students who live off-campus.

Until recently, when you walked into the fifth floor office of the Palo Alto Building Department, the first sight that greeted you was row after row of bright pink "Stop Work" notices taped conspicuously to the wall next to the counter. Although these are now stored out of sight, these served as public notice that a lot of people were doing a great deal of improper and illegal building. The greatest number of Stop Notices were issued for "Failure to obtain a building permit." For the part of me that takes sadistic pleasure in reading about other people's worries and woes, these notices were great fun: improper setbacks, not in conformance with plans, installation of illegal materials, houses built on the wrong portion of the lot, houses built on the wrong lot altogether! And on they would go, ad nauseum.

In Palo Alto, and in most towns and cities, the building permit law is very simple: if you're doing any building at all, you need a permit. To obtain one, you either have to be the property owner or a licensed engineer, architect or contractor. If you are one of the three latter, then you must have a Worker's Compensation Insurance certificate on file, or else you must sign a statement that you will not be using any employees on the job. Although not required by the city for a permit, it's more than a good idea to have liability insurance.

Before I became a licensed contractor, I never bothered with permits, even though one was legally required for the small jobs I would take on evenings and weekends. When I finally obtained my contractor's license, applying for that initial building permit was a first class encounter with high anxiety.

I walk into the building department and I am immediately confronted by the wall papered with all those bright pink Stop Work notices. The next thing I see is the back of a huge computer screen, behind which towers a wide, dour looking guy standing about six feet four and wearing his green Inspector's Badge.

"Here for a Building Permit? Fill this out," he says and hands me a permit application form.

I take a long time to read it over, and don't understand half of the information they are asking me for. When I think I'm done I hand it back to him and he quickly feeds my information into the computer.

"Is this your right license number?" he asks abruptly.

"4-3-8-0-9-9?"

"Well, I can't give you a permit. This license is suspended," he replies, looking even more dour than before.

"That can't be. I just got this license!"

"Just kidding," he says, now with a dour-looking smile.

I smile back faintly, feeling more relieved than he knows.

"The truth is it's not in here," he says after three more computer clicks. "Ever done work in Palo Alto before?"

"Not legally," I almost confess. "Not recently," I say instead.

"Let me see your license then, and I'll put you in the computer. You do have Workers Compensation Insurance, don't you?"

"I sure do," I lie, as I proudly hand him over my brand new Contractor's License. I haven't even folded it yet.

"Okay. Then call your carrier and have them send us a confirmation notice. Who is it? State Fund?"

"Yep!" Who is State Fund?

"Well, they're good. They'll send it to us the same day you call them. You can use the phone on the desk if you want."

I don't want.

"Thanks," I tell him. "But I'll have my secretary call them from the office."

It's a week before I actually do get the worker's compensation insurance. When I do come back, thankfully someone else is behind the counter.

Over the years since then, I've had all of the Palo Alto Building Officials come out to inspect one job of mine or another. There are only two don't I like: Inspector Frank and Inspector Phillipe. From both I get the feeling they hate their job. Every time they come out for an inspection, they're very sombre and sullen, and while 90% of the time they sign off the job Inspection Record, they always have negative criticisms to make. In checking with other contractor's, I find these two guys operate the same way on all their jobs.

One day at lunch I run into Inspector Bob and I ask him about the different inspectors in the department. Inspector Frank, I learn, has passed qualifying tests in every single area of inspection, from major skyscraper construction to specialty solar installations. Inspector Phillipe is an immigrant from another country where he was a licensed engineer. However, in America, he has been unable to obtain engineering work. Both men feel overqualified for their inspector job and resent not having other work.

It's no surprise these guys can't get other work. They

don't appreciate the work they do have. Even if they got jobs that matched their level of knowledge and experience, once the romance wore off, they'd soon be feeling resentful all over again. I'd like to encourage them to try psychological counseling. I'd probably be doing the whole City a service.

Still, I recall the times I've been unhappy in my own work. Marsha Sinetar, a career counselor, believes that work and unhappiness do not have to go hand in hand. More is possible. I can experience a sense of vocational integration:

"The individual who achieves vocational integration is able to fully focus on what he is at the core. . . . Work is one of the ways that the mature person cares for himself and others . . . Moreover he is able to be that unique, distinctive personality in and through his work because he has gained a specific, vital, unobstructed strength which empowers him, which is life-giving, energizing, creative, loving. This added power is gained when the individual sees himself as he really is, understands what drives him, what motivates him to act — or to back away from things — when he knows what he wants to live for (perhaps even to die for), and then consciously chooses to enact that knowledge in everything he does. Naturally, his work falls into place as the correct thing for him to be doing (or, when he senses it is not appropriate for him, the incorrect thing.)

The vocationally integrated person is able to choose to risk all and surrender to the truth of himself as he exists within himself in that most private and sacred core of self, and this choice . . . is the beginning of the dissolution of his limited, egocentric former life. However, it is this choice which initially produces the profound risk and death-related images and emotions that are so troubling. Happily, once the initial steps are taken to settle for nothing less than living out the truth and reality of his life, other lesser choices become easier and the actualization process (and with it vocational integration) is freed to accelerate."[1]

Inspector Dan rightfully deserves the job of Supervisor of the Palo Alto Building Officials. He has the "vocationally integrated" attitude that Building Inspectors are public servants, very similar to police officers. While it is his job to enforce the building codes and city ordinances, his real job is people-serving. When Inspector Dan walks onto the job, it is not as an authoritarian adversary, but as a collaborative ally, available to help produce the best job possible.

Many building officials come to their job from backgrounds in contracting and construction. Just as few contractors know everything in the National Plumbing, Electric, Mechanical and Building Codes, neither do most building inspectors. That's why they carry the code books with them. But they do have considerable general knowledge, and their inspector's experiences to guide them. They must also obtain continuing education and training.

Once I obtained my license it became clear to me that building inspectors and I were on the same side of the fence. There are 82857 licensed general contractors in California.[2] That breaks down to one out of every 333 persons. If you include all the licensed subcontractors and licensed engineering contractors, the ratio is one out of every 275. Then add in the people who do work without licenses, the ratio becomes something more like one out of every 250 persons. That's a lot of competition. Because of building inspections and permit requirements, I know that 9 times out of 10 when I give a competitive bid on a construction work, I will be competing against other licensed, insured contractors who will be required to obtain the same permits, carry the same overhead costs, and perform up to the same level of competence as I will. In this way, Building Departments and Officials operate for the benefit of my best interests.

There are times though, when interests inevitably conflict. The majority of my conflicts with building departments have come with rookie inspectors. Such a conflict occurred with New House's drywall inspection.

Before tape and spackle can be applied to drywall, an inspection must be made to insure that sufficient nails or screws have been installed, that all corners have proper backing and that the proper fire rated boards have been placed where required. On Tuesday, I call for an inspection of New House's drywall nailing for Wednesday.

Early Wednesday morning the familiar white car with the official green emblem pulls up to the job but a totally unfamiliar figure steps out. Its a thin young man, about 22 years old, with a sport coat and a clipboard.

"Good morning. I'm Inspector Tim. You requested an inspection for drywall nailing?"

I nod.

"May I have your permit and inspection record please?" he asks.

Whenever anyone calls for a building inspection, the person who takes the call logs it into a log book and simultaneous onto a six by eight inch carbonized paper form which lists the building address, the contractor, permit number and type of inspection requested. After the inspection is made the inspector simply writes "pass" or the corrections he wants made on the form and gives the contractor a copy. Even Frank and Phillipe do that. My New House file folder has about 30 of those sheets in it. I'm beginning to suspect that Inspector Tim is somewhat new at this work.

"My building permit and inspection record?" I repeat back as if he's asking for some strange foreign document. "Hmmm. I must have left it at the office. Can't you . . ."

"The law requires that the permit and inspection record be posted in a conspicuous place and be available on the job at all times. I'll let you off with a warning this time."

"Thank you. I'll be sure to have it here next time," I tell him, using every power I possess to suppress the sarcasm and disdain I'm feeling.

"Is this fire-rated type X sheet rock here in the garage?"

"Yes it is."

"How about under the stairway?"

"Same thing there," I tell him.

"How come I don't see any markings indicating that?" he asks.

"The markings are put on the ends by this manufacturer. Anyway you can check an edge at one of the window openings and see a marking there."

"Alright, I believe you. But what about this nail spacing?" he replies, quickly changing the subject.

"What about it? Three through the field and one at the edges," I respond. An edge is now clearly in my voice.

"Code requires twelve inches on center maximum. Some of these are 10 and 11 inches on center and some are 13 and 14 inches. There's no way I can pass this job."

"Fine," I tell him abruptly. "We'll fix it and call for reinspection tomorrow. Thank you and good day."

When Inspector Tim leaves, my first thought is, "Now that guy was a stupid ass". Some I attribute to inexperience. But inexperience is one thing. Asinineness is another. I think the city ought to shape up their Inspector training program. That guy's lucky I was his first inspection job. I know some contractors who would have "accidently" dropped a glulam on his head. They probably decided to break him in on me because they know I'm such a kind and gentle, non-threatening soul.

Obviously Tim was correct in the letter of the law, but he was stupid in the spirit. He's obviously never nailed up drywall. When you're holding a 70 pound, twelve foot board over your head, you're not thinking about a knat's ass variation in the nailing pattern. So in this case, it's not only inexperience, it's ignorance. He's never done any building. He doesn't know what's practical.

The next morning the same white car drives up, but Inspector Dan steps out. I walk over to the car and he hands me an inspection ticket with the word "Passed" written across it.

"Don't you even want to look at it?" I ask.

"You told Tim you were going to fix it, didn't you?"

"I didn't really think it needed fixing, but I did send somebody through to double nail wide spacings."

"Well, look," Dan says. "He's new. He's just starting out and he's got to start some place. He's going to make mistakes. Just like me and just like you. By the way, are those aluminum heating ducts I see hanging down there in the garage?"

"I guess so. The mechanical man put them in."

"Shame on him then. You can't use aluminum exposed in the garage. It's not one hour fire-rated. He ought to know better than that," Dan says, smiling. "You should too. How long have you been in this business?"

"I get the point," I say, smiling back. "I'll take care of it. And feel free to send Inspector Tim back when you think I'm ready."

This incident with Inspector Tim is almost pleasant when I consider the time it took five inspections to get the installation and nailing of plywood roof sheathing signed off on another earlier house.

Installing and nailing sheathing is pretty fundamental. On this house I leave Ramon to install the plywood, check the nailing and call for inspection when he's ready. The day the inspector comes out I'm not at the job. When I do return, I find Ramon visibly upset.

"What happened?" I ask.

"The inspector wouldn't sign off our sheathing because we used staples and the plan called for nails. I even showed him the letter from the engineer okaying staples. He said he wants the engineer to call him."

"Okay," I say. "We'll have him call. No big deal."

The engineer calls and apparently the inspector is satisfied. We reschedule inspection number two.

Other work calls me away and I miss the inspection again. When I return, Ramon greets me with more bad news.

"It's the same inspector, but he's really new at this," he says. "He wouldn't sign it off this time because he found a half dozen places at the edges where the staples were spaced more than four inches apart. Plus he wants certification for the staples we used."

"Did you tell him you'd nail the plywood while he was here?"

"He wouldn't wait. And he wouldn't inspect any more of the roof. He told me to inspect the whole roof and call for reinspection."

"This guy is very strange," I say. "he must think they pay him by the piece. What's his name?"

"Percy."

"What?"

"Inspector Percy. Percival Plumbchuck. Something like that."

"Geez. No wonder he won't pass anything. I wouldn't either if I had to carry around a moniker like 'Percival!'"

Three days later, which is as soon as he's available to return, Inspector Percy performs another inspection. I'm there for this one. He climbs up on the roof, looks around and then stops.

"This plywood is supposed to be CDX," he says. "I don't see any markings saying that on these sheets."

I proceed to point out the C's and D's and the words "exterior glue" on a number of the sheets.

"What about these sheets with no markings," he asks seriously.

"Those are probably nailed with the markings face down," I tell him as patiently as I can.

"Well, let's go down and check."

Down we go to check and sure enough, a number of the sheets have the markings facing down into the house.

"Well, there's at least three sheets I noted on the roof that don't have markings on this side either," he says and proceeds to put red X's on three plywood panels.

"I'm afraid I can't sign this off."

Since I got all the plywood from the same yard, who got it all from the same mill, quite frankly I cannot think quickly enough to come up with an explanation to keep him from driving off. At this point I'm beginning to believe he does think he's paid by the inspection. But I have real cause for concern the next time he shows up.

Each of the sheets Percy marked with an X turns out to be less than a full sheet. The portion with the identifying markings has been cut off and used somewhere else. I call him back for inspection number four. He doesn't even go up on the roof.

"When I was driving away last time I saw a number of pieces of plywood that were stitched together with string. I couldn't find anything in the code book where it approves this kind of plywood. So I'll need you to get some kind of acceptable documentation before I can sign off on this."

"Why even bother to come out then. You have the phone number here at the job." I am speaking with barely controlled anger.

"I came out because there's something else I noticed

that I'm not going to pass. None of these 2 x 4's have any nails in them. I don't see a single nail. They're not nailed to the plates." he says, pointing around to our top and bottom plates. Suddenly, I am struck with the notion that this is not a real inspector at all. Inspector Percy is an actor the guys have hired to pull a practical joke on me. I begin to crack a smile until he pulls out an official pad of correction notices and starts writing.

"Wait a minute. Are you serious?" My voice is very loud. "These studs are nailed. They're nailed on the floor through the bottom. We raise the walls into place afterward."

Percy pauses for a minute to let this information sink in. Then he says, "But if I can't see the nails, how can I inspect them?"

"Look, I'll get the documentation for the plywood. The nailing's fine. Just go back and ask someone how to inspect framing nailing, and we'll call you back when we're ready. Til then, have a nice day and we'll see you later." I can hear my own words being spoken very slowly and filled with menace.

Two days later, after our fifth inspection, Percy returns and finally signs off on our roof sheathing. He makes no mention of our stud nailing. And I make no mention of his ignorance.

After incidents like the two above, as an investigative exercise, it is often helpful to ask: Why were these people sent to me?, or better, Why did I attract them? What lesson can be learned?

My first response is they are present as teachers offering me a chance to practice patience in the presence of people in authority roles who have less knowledge and experience than I do. True knowledge and experience are the only real authorities. Recasting them in different roles I can easily imagine they have shown up on the job to test my own internal psycho-spiritual quality control. These two times I rightly failed to get signed off. Perhaps I'll pass in the future on the forth or fifth inspection.

14

"By the work one knows the workman."
— La Fontaine

Defining Interior Boundaries

Until the drywall goes on, New House remains internally open and allows people to look easily throughout the whole structure. It's an interesting experience. You can look from bedroom to bedroom, bedroom to bath. From the kitchen you can look directly up into the study or straight across through the bathroom into the family room or the dining room. It's a lot like standing inside the skeleton of a large, living organism. Without the walls to create boundaries, there is a lack of definition and clarity. Walls create the edges and limits that define how the house is used — ultimately, its character. We are similar: our edges help define our character

With all the internal organs (wiring and plumbing) and the skeleton (framing) in place, it is time for the internal skin — for the drywall to be hung, for New House's boundaries to be further defined. Stan has recommended that I use Richard, whom he knows from other jobs, for hanging and texturing our drywall. Richard's price is somewhat higher than the other bids I have received but, if his work is good, then the budget can afford it.

However, after Richard's first week on the job, I am ready to jump off the roof. First, he never stops talking. All day long, yak, yak, yak, yak, yak. And all the while he's talking he's wearing an earphone radio with music playing so loud we can hear it through those little speakers when we're downstairs and he's upstairs!

Second, he's an erratic worker and on the phone constantly. Richard comes to work at 9 AM. and, by the time he's unpacked and ready to work its 9:30. Ten o'clock is break time. 11:30 is lunch. What's more, during that time he's made at least a half dozen telephone calls. After lunch he's on the phone for a half hour to an hour at a time with his ex-mother-in-law, his girlfriends, his customers, his attorney, his dead wife . . . And then people start calling him on the job . . .

His dead wife? Yeah. She's been dead eight years and that's all he talks about. Not about her being dead. But about how he's unable to collect the money from his lawsuit, how his mother-in-law is trying to screw him out of his share, and on and on ad nauseum.

Finally I simply tell Richard not to use the phone. I tell

him over and over and over, but he doesn't listen. He doesn't get the message. I check with the phone company for a two week total of his charges and it's over a hundred dollars! This situation is out of control and must be changed. I can't take the phone out because I need it. I consider telling Richard to take a walk, except he *is* a good drywall finisher. He's just *not* a good worker. And besides, it's a hassle to try to get somebody else to finish up a job that's half started. Few people are willing to walk in and clean up another person's mess.

I'm at a loss over what to do. Mitch, Carlos and Ramon continually here me whine, moan, bitch and complain. To get the situation in hand, I finally resolve to treat Richard like a misbehaving child. Although it feels disrespectful to treat a grown man like a child, that's how he acts.

On Monday I replace the pushbutton telephone with a rotary dial and install a lock on it. I inform Richard that his time on the phone is causing me to miss important calls and that he should only use the phone for life and death emergencies. I tell him to save up his calls and make them on a pay phone at lunch time, off the job. Lastly, I tell him if the people he needs to call are out during lunch, then he should leave early, or arrive later, and call them when they are available. End of discussion.

For two days Richard sulks and works very slowly. Even though he's working by contract and not by the hour, his behavior is clearly intended as a silly form of revenge. When he comes even later and then leaves even earlier than before, I finally decide I'm no longer willing to put up with his behavior. I point out the portion of our contract which imposes a $100 per day penalty for delays caused without excuse. After much whining and moaning, Richard stomps angrily off the job. The next day he shows up at 7AM with two other tapers and completes the remainder of the work in three swift and silent days. Goodbye and good riddance, Richard!

Later I realize that Richard is not fully departed. All the negative thoughts and feelings that arose out of his presence on the job still remain with me at New House. With him physically absent, I more easily see these thoughts and feelings are mine! They have always been mine, are still mine, and serve as good markers for my own internal boundaries, those places that are my growing edges.

Just as New House has internal boundaries, so do I. People like Richard, who aggravate and frustrate, bring those edges into sharp focus. Like Brenda and Homer, Richard is a teacher. (That doesn't mutually exclude him from also being an ass.) He brings those areas into focus where I could use more development and awareness including learning to express myself directly, with less inhibition, confronting conflict head-on, and trusting I can handle whatever problems my honest, direct actions may create.

With the help of Richard I can more readily see internal boundaries being stretched and opened by the building of New House. Doing this work expands me professionally. I am able to provide more and more people with more and more useful work. I am required to trust that the men who work with me know what needs to be done and will do it, which is another internal boundary I am attempting to stretch. I am learning to trust that the crew working on New House can, and will, build it within budget, and within a reasonable timeframe. It is certain that the men will make mistakes. I cannot recall how many mistakes I have made in my career as a builder. But I am fortunate in remembering that being condemning and critical serves no good purpose. Making someone else wrong usually only prolongs the period of time *I* remain upset. The most effective way to correct mistakes is by giving clear and concise information. When I deal with the mistakes of others this way, I find that I notice and respond to my own mistakes earlier, and in a clearer and more self-forgiving manner.

Threaded through all of this work — through all the observation, examination, and investigation, through the attempts to identify personal boundaries that can be stretched, old behaviors that can be dropped, new behaviors that can be learned, through all the changes that contribute to me growing as a builder — threaded through it all is a belief that there is something different and better to become. My friend Duane suggests it is to become a Conscious Housebuilder and offers me a list of such a housebuilder's characteristics.

Strengths of a Conscious Builder

A conscious builder is a strong builder. He is able to stand back and look at a variety of choices as well as observe himself in the choosing process. He is able to objectively appreciate his own unique character, sentiments, and goals. A conscious builder has backbone, courage and determination. Some other strengths that characterize a conscious builder are:

- Informed — A conscious builder is well-informed. Instead of moving through life half-asleep and unaware of the challenges he faces, a conscious builder is disciplined in learning about important trends and issues.
- Self-reflective — A conscious builder knows his own mind and does not blindly trust authority, experience or ideology.
- Courageous — Because a conscious builder knows his own mind, he can move ahead with confidence and assurance.
- Forgiving — A conscious builder recognizes that learning inevitably involves making mistakes. Therefore, errors are not automatically defined as being "bad"; rather they are embraced as useful feedback and recognized as grist for the mill in the process of learning.
- Dispassionate — A conscious builder is objective, cool-headed and reacts calmly to the stressful pushes and pulls of trends and events. He has acquired an evenness and level-headedness that is not pulled off-center by the passions of the moment.
- Inclusive — A conscious builder hires and works with all peoples, regardless of race, creed, color or ideological perspective.
- Anticipatory — Because a conscious builder stands back to view the work more objectively, he easily explores how current choices and decisions will impact future work.

- Experimental — A conscious builder is not locked into habitual patterns of thinking and behaving. He does not respond to challenges with preprogrammed solutions, but engages situations with freshness and flexibility.
- Responsive — A conscious builder does not wait passively until some crisis demands action; instead, he is able to respond in a swift and timely fashion by mobilizing an informed and attentive awareness.
- Self-orienting — A conscious builder continuously re-orients himself by cutting through distractions to the core of issues and problems.[1]

Duane is probably right. These are directions in which I am attempting to stretch my boundaries. But ironically, the personal area that causes me gravest daily concern — where I consider myself least conscious — is my pattern of eating. Because I am able to exert so little conscious control over my eating, a benefit of being in the building business is that I expend so much energy at work that I can eat whatever I want, in any quantities I want, and not become grossly obese. I've been consuming large amounts of sugar daily since I was five years old and, while I've taken a number of medical tests to discover any adverse health consequences, nothing has ever turned up.

I have no illusions but that my craving for sugar is an addiction. Unfortunately, my addiction is socially acceptable and, more often than not, something to laugh and make jokes about. For me it's mostly something not to think about — something to be controlled by, and to harbor hurts and resentments about. The only times I've ever been able to exhibit even a small amount of conscious control over the craving for sugar has been short periods when I've deliberately abstained from all foods — times when I've been doing a cleansing fast.

The countless errands and trips during the building of New House adds to the problem. Instead of packing a lunch, I usually stop at a fast food restaurant and pick up something sweet and greasy to go. Then too, the large hardware stores and lumber yards I visit on almost a daily basis have candy and snacks available to the customers, usually right up front near the cash registers. As I stand in line drooling over the sweets, a part of me knows that there is definitely a conspiracy at work. The world knows I have a problem I can't control and they'll take every advantage in order to make a buck.

Because I am unable to stop or change this habit (or start a new one), I banish it from all conversation and do my best not to think about it. My efforts at deinal meet with futility and frustration, for the problem has its own life. As the song lyric goes,"I can never get away from me."

I have many justifications for my lack of success in the area of diet and nutrition. One is, "I'm simply grabbing for all the gusto that I can." Another is, "It takes a big man to do a big job.' (And not, "It takes a big, fat, slob to do a slob of a job.") The rationalization that offers the most solace is: "Buddha had a belly. It's one of the require- ments to be come a great spiritual teacher." But still the discomfort persists. The belt is perpetually a little too tight. The ladders just a little too hard to climb. The path just a little too far to walk and the lesson just a little too hard to learn. Except that this class is never ever really dismissed until the last day, when school is out. Still, I know I can keep working for that degree.

15

"The biggest source of human suffering is
the simple thing of letting our mind
wander away from its home."
— Marian Mountain

Break Time

A flurry of New House activity is scheduled for one
Friday: the driveway and sidewalk are to be poured, the
hardwood flooring is going to be laid throughout, the
kitchen appliances are coming, the carpeting is going in
upstairs, and the burglar alarm crew will be connecting
and closing up all their wiring. And this is also the day I
have decided to spend golfing.

More specifically, it is the Friday that I have been invited
by a friend to attend the U.S. Open Golf Tournament at
the Olympic Club in San Francisco. My interest in golf
began a year ago, after reading Michael Murphy's book
Golf in the Kingdom. I worked up the courage to try and
play the game and found it enormously challenging. Since

then, I try to play at least two or three times a month. But running off to watch other people play seems like a dereliction of duty.

The decision is truly a difficult one, particularly with so much activity scheduled to take place while I'm away. I feel guilty having fun while all the other guys are busy working. But I'm entitled to have a break every now and then. I mean, it's not like I don't think about New House 16 and 18 hours a day. Besides, how many U.S. Open Golf Tournaments am I going to have the opportunity to see first hand in my life. I clearly outline and delegate responsibilities among Ramon & Mitch, give up the guilt, and then go fully prepared to enjoy myself.

I spend the day at the Olympic Club and walk the fairways with Jack Nicholas, Tom Watson, Greg Norman and a whole host of other golfing greats. It is an extraordinary day. These men exude an electrifying spirit and presence — they are enlightening simply to be around. The daily work of housebuilding seems to pale in comparison. I'm soon to learn there is a high price to pay for enlightenment.

The next day I discover that, in my absence, no piping has been laid under the driveway for future sprinklers or electric wiring. Now pipe must be run completely around the house, resulting in additional labor and material costs. I feel guilty and frustrated about this situation.

An even more serious problem arises on Saturday — Mitch snaps an Achilles tendon during a tennis game. He undergoes an operation to have it sewn back together on Sunday and will be in a full leg cast and out of commission for at least eight weeks. During that time all the crucial trim and finish work is scheduled to be completed on New House. I am tempted to search for a connection between my going off to watch the golf tournament and Mitch's injury. Is this a proper penalty for improper behavior?

It is difficult for me not to feel both victimized, and at fault by what has taken place. While there is nothing I could have done to prevent Mitch's injury (he didn't even get hurt on the job), a big part of me simply doesn't believe in accidents. As a keen student of unconscious process, I wonder about the implications of the incident and what they represent. The best I come up with is that the break in Mitch's Achilles tendon represents his need to take a break and discontinue building on New House. I wonder if being fully responsible in my absence on Friday caused him to literally snap under the strain on Saturday.

I ruminate on cause and effect, how events that happen on Friday impact next Saturday and every day thereafter, and how little I am able to comprehend all the ramifications, particularly as my judging mind attempts to classify events as good or bad. A familiar story helps me put my concerns to rest:

> "A farmer who had just acquired a stallion came to a Zen Master in distress, saying, 'Master, the horse is gone, the horse is gone!' for the stallion had run away. The Zen Master replied, 'Who knows if it's good or bad?' The farmer returned to his work feeling sad and miserable. Two days later the stallion returned and brought with him two mares. The farmer was overjoyed and he went to the Zen Master, saying,

'The horse is back and has brought two others with him.' The Master replied, 'Who knows if it's good or bad?' Three days later the farmer was back again, crying because his only son, his only helper on the farm, had been thrown from one of the horses, and his back had been broken. He was now in a body cast and could do no work. The Zen Master again replied, 'Who knows if it's good or bad?' A few days later a group of soldiers came to the farm as they were conscripting all the young men in the area to fight in a war. Since the farmer's son was in a body cast they did not take him."[1]

With this story in mind, I realize I truly don't know if the incidents that occurred in the wake of my U.S. Open holiday are good or bad. I resist the temptation to blame myself and instead consider the best plan of action to move forward. I have two clear options: I can do the work required to finish New House myself, together with Ramon and Carlos, or I can advertise for an experienced trim carpenter to help do the finish work in Mitch's place. I don't like the thought of bringing in a new, unknown person onto the job, and anyway, first-rate help is not usually available on a pinch-hit basis. Given these considerations, I elect to postpone the decision for at least a week.

In that time I talk to Mitch and get his ideas about what to do. He confirms that he will not be available, and apologizes for the inconvenience. He also offers me the names of several friends who are proficient at finish carpentry. I call one of them, and he's willing to come and work for a month. He shows up in a Volkswagen Beetle on a Monday, late, hungover, with no tools. I tell him thanks, but no thanks. The wisdom of my plan to take a break for a week before making a decision has just been confirmed.

Thinking about taking that break reminds me of how I have learned to approach smaller breaks during the regular workday. From my first days in construction, I have seldom taken the customary fifteen minute mid-morning and mid-afternoon breaks on my jobs. Initially, I found the work so invigorating that I did not want to stop arbitrarily at 10:15. It broke the flow — the rhythm of the work — so I would simply work on. This approach made me less than popular with those who spent more time looking at their watches than at their rules and tools. But the barbs and jibes were usually good-natured and addressed to me as the quintessential oddfellow, comments that I have grown accustomed to over the years.

My lack of interest in taking such breaks is usually accepted as the modus operandi by my crews. Surprisingly, no comment has ever been made about it. At New House we work until lunch and, after lunch, we work through until it's time to pick up the tools at the end of the day. When other subcontractors come onto the job, most of them will take the morning and afternoon coffee break while my guys will work right on, usually thoroughly absorbed in their work. Often if a subcontractor is on the job for more than a week, he will give in to the group dynamic and begin to go without his own customary break.

But my real resistance to establishing routine breaks

evolved out of those early experiences as an apprentice. Customarily it is the foreman's responsibility to notify the crew when it's time for break, but invariably on those jobs, as break time would approach, some other crew-member would shout out "break time," thus ursurping the foreman's authority. Even though I continued to work, I always sensed these incidents had a hostile aspect to them. Once the break was underway, normal chatter and horseplay would take place until just before it was time to go back to work. Then someone (usually the same guy who shouted "break time") would start to tell a long, involved story obviously designed to extend the break a good five or ten minutes. As I observed the push/pull adversarial nature of the relationship between the foreman and the crew, somewhere along the line I decided it would be easier to forego the stress of those dynamics and skip breaks altogether. I would only deal with them once a day — at lunch.

That's not to suggest that I never take breaks as I work on New House. I do. It's just that I take them when the need arises in the natural course of the work. When a job is completed and it's time to move onto the next piece of work — that feels like the best time to take a break. Other good break times are when a piece of work is going badly and I need to step back and get some perspective.

As my work has moved away from the day to day hammering and nailing, the primary motivation for breaks has changed from being frustration-driven to anxiety-motivated. My frustration with the actual hands-on work (such as trim I just can't cut right, or plumbing fixtures that won't stop leaking) has given way to anxieties about how much there is to do and how will I ever get it all done and paid for.

Anxieties also arise when I perceive that things are out of control. I have learned that in losing control, what I have lost is my ability to start, change and stop things. Such a loss can be compared to a mass of calls coming into a telephone switchboard simultaneously. Each call insistently demands the attention of the operator. Control is begun to be regained by the operater stopping just one demand. It doesn't particularly matter which demand is stopped. Handling just one call permits handling another and so forth until the condition of the switchboard is changed from total confusion to a controlled situation. Confusion and loss of control take place when there is nothing in a situation which I can stop. When I can at least stop one thing in a situation, I then find it is possible to stop others. Eventually, I will recover the ability to change certain factors in the situation. From this I gain an ability to change anything in the situation, until finally I am capable of starting my own line of action.

An avalanche of people and things need my attention at New House on a regular basis. They are much like the calls coming into the switchboard. Because I am responsible for the total project, I must pay attention to every aspect of the job. An inordinate amount of my time is spent on planning and coordination, leaving little time to put on the nailbags and do actual construction. Still, I deliberately try to do some bit of physical work every day to act as an anchor. When there seems to be little I can stop, start or change, when confusion reigns and things seem out of control, the ability to pick one task, focus on it, and see it through to completion is the first step in making my way out of the jungle of chaos.

To complete the planning and building of a house requires many months of sustained energy and focus. Upon completion, I need a break to restore my energy and psyche. Although I have built more than two dozen houses over the years, I've never built more than three houses consecutively without large amounts of time in between. Without deliberately planning it that way, my housebuilding career has had significant break periods interspersed throughout. I took several years off to complete college degrees. One period of time was taken to learn my way around a woodworking/cabinet shop. Other breaks have been taken simply to vacation and relax. My latest break is the time being taken to complete this book.

After a long period away from housebuilding, I am able to come back to it renewed, with a fresh sense of the work. Then, when I walk back onto a building site, each project feels new — as if I am coming to work for the very first time all over again. There is an acute sense of anticipation and excitement circling around in the blood and the brain. I feel both light and invigorated at the same time. Getting back to work is then like coming home anew after getting all the breaks.

16

"Caress the divine details."
— Vladimir Nabokov

Trimmed to Perfection

The quality of construction that goes into a house is most clearly evident in the time and energy spent on the interior trim. Woodworking joints must fit together tightly, doors must sit equally spaced in their jambs, grout lines on floors and vanity tops must be proportional and straight, paints and caulkings must be smooth and uniform, and a whole host of other details must be able to withstand inspection from my own laser-like eyes.

Knowing how the interior finish should look and actually accomplishing it are two very different things. I was counting on Mitch, to help me achieve the quality of interior perfection I want New House to demonstrate. As

I learned upon returning from the U. S. Open, Mitch has snapped his Achilles tendon and will be on crutches and in a cast for eight weeks — the time within which all the trimwork on New House must be completely finished.

The week of reflection, and the walking disaster Mitch's friend turned out to be, still haven't helped me to make a decision about how to handle the trim. I probably should advertise for someone else right away to take Mitch's place, but I have a number of conflicts. What happens with Mitch when he's healed? What do I do with the new guy? Suppose he turns out to be better and faster?

Should I do the work myself or try to get Carlos and Ramon to do work they are not experienced or equipped to do? Realistically, I'm not that proficient at trimwork myself. If I relay on Carlos and Ramon alone, the quality of the work will most likely suffer and so will the quality of my emotional well-being.

I decide to talk to my wife and call some friends for advice. Based on their input I decide on a combination of all the options: I will let Ramon and Carlos attempt what I think they can handle reasonably well, closet base, shelves, poles and other simple things. I will install trimwork that I enjoy and am relatively good at: hardware, window sills and aprons, and finish plumbing fixtures. Finally, I will get someone who is proficient at doors, casings, and all the other intricate trimwork that New House requires. This will likely be someone who is currently holding a full-time job and can moonlight evenings and weekends.

Trimwork done well goes slowly. It is difficult to achieve the degree of perfection that both the customer and I would like. I strive to achieve an ever greater degree of perfection. I encourage the crew to take the "Japanese approach" and make a commitment to tiny improvements in a thousand places. I feel as if increasing the level of craftsmanship is a drive that has been genetically encoded in us building creatures from the moment we were first conceived.

Parallel to the quest for increasing levels of perfection in New House is my own quest to increase the level of my interior perfection. This inner quest is connected to my desire to stretch interior boundaries. As I age and change, certain things that used to be of concern, like financial and social security, no longer have the power over me they once did. There was a period of time in my life when those things were of driving concern, and I worked to achieve a high level of material success that felt appropriate and matched my needs as I perceived them. Now it is nearing time to respond to the faint cry of other needs wanting a larger voice. Millard Fuller, hearing a similar voice, quit his work and founded the non-profit housebuilding organization, Habitat for Humanity. Here is how he describes his journey:

"From the first day of our partnership my partner and I shared one overriding purpose; to make a pile of money. We were not particular about how we did it; we just wanted to be independently rich. During the eight years that we worked together, we never wavered in that resolve. And when the treasurer of our company walked into my office one day in 1964 to inform me that I was worth a million dollars, it came as no great surprise. I accepted her report with satisfaction and turned immediately to my next goal: *ten* million dollars.

"But everything has a price. And I paid for my success in several ways. One high price I paid for my personal affluence was a compromising of my personal morality and integrity . . . Another big payment was losing my health. I developed a severe breathing problem. Frequently my chest felt as if tremendous pressure was bearing down on it. Often I had to leave my desk and walk around the office or even around the building in order to relax enough to resume normal breathing. Then my wife precipitated a crisis . . . She left me.

This, I discovered, was a price I was not willing to pay. We separated for a time and when we began our time of reconciliation, my momentous decision was already beginning to be made. When she agreed with me, I knew it was absolutely right. We would sell our land and houses and boats and cars and cattle and horses. We would also sell the business-to my partner if he wanted it, and to someone else if he didn't. *And we would give all the money away.*

We had gone too far down the wrong road to be able to correct our direction with a slight detour. We simply had to go back and start all over again.

Together, my wife and I embarked on a tremendous journey of faith.[1]

Fuller's conversion was sparked by personal crisis and his involvement with the Christian religious tradition.

In some respects my own flight into housebuilding away from the manufacturing business I owned, is similar to Fuller's. My move was inspired in part by poor psychological and physical health, but it too, was a journey of faith. On such journeys the expectation or hope is that things will perpetually go perfectly.

Such hopes are best abandoned, for crisis quite frequently acts as a catalyst instrumental in nudging us along the path of growth — although it's not a requirement for development, nor is it my personal path of preference. Abraham Maslow, the father of humanistic psychology, described major stages of growth that happy, successful people seem to move through on their way to maturity. Briefly, it looks like this:

- Basic physiological needs (air, water, food)
- Safety and security needs (shelter, health, protection)
- Love and belongingness (family and community)
- Self-esteem (self-approval, self-validation)
- Self-actualization (an integrated, balanced, growth-oriented person)[2]

Drawing from my own life, I do know I can meet my own survival needs — the first two stages (that include air, food, water, shelter and clothing). Air and water are generally free, and to my wife's chagrin, clothing has never been a real concern. Astonishingly though, I am only now beginning to deeply feel that I have the physical, intellectual, emotional and psychological capacity to meet those needs in the course of everyday living. I also feel a part of a community of family, friends, neighbors, associates, and just people in general — all of whom are part of a supportive network that will always be in positive relationship to me, and me to them. Through this community I feel my needs for food and shelter will always be met. My ability to provide for these needs has always been present. The problem was I did not possess the sense of self-esteem and self-confidence that allowed me to know with certainty, I could provide for those needs.

But now survival and security needs no longer require so large a portion of my daily attention. The focus and emphasis of my interior concern is moving a little bit farther up the hierarchy. The universe gives powerful opportunities to learn where my real sense of safety and security is to be found. Shortly after my wife, three-year-old daughter and I move into our new neighborhood, the next door neighbor's house is burglarized. The burglar used a prybar to force open a side window in the middle of the day and walk off with the neighbor's VCR and several cameras. The police explain that in recent weeks over a half dozen burglaries have occurred in a ten block radius, all during the day, and all with the same method of operation. This news does not do much to quell my ever-present storms of paranoia. I feel particularly unquelled when, three days later, the same guy comes back to the neighbor's a second time and walks off with his TV and stereo.

I am determined to keep him out of our house. That weekend I install deadbolt locks on all the doors and put wooden blocks above all the lower halves of the double hung windows. I close off access through the crawl space and I put torque head screws into the frame to secure our skylight.

Three days later, it is four in the afternoon, time to pick up my daughter from daycare. I come home to clean up before taking her out to dinner and watch a little of the Giants baseball game on TV as I get ready. I then close everything up and leave. After I pick up Amanda I realize I have left my wallet on the bedroom dresser and I must return home for it. I have been gone only about twenty minutes. We walk in together and I go into the bedroom for my wallet. I pick it up and as I walk back into the living room, I can't believe my eyes: a bunch of wires are dangling over the cabinet where the TV was playing the baseball game only a few minutes ago. The sense of fear and outrage is enormous. I immediately sense that too little time has passed for the burglar to get in and get out with my TV, so I begin looking for the guy. I open closets and look in rooms and cabinets, and then I stop . . . I remember Amanda!

I rush out to the living room, take her, and immediately leave the house. We get into the truck and then drive straight to the police station several blocks away. We tell the police the burglary may still be in progress. They send a car over and, sure enough, there's a guy walking down the driveway with my TV in his arms. He throws the TV into the bushes and takes off, but the police catch him. It turns out he's armed with a hunting knife, several screw drivers and a prybar. Later I discover he ran and hid with the TV in our laundry room closet when I came home with my daughter. He has several assault and burglary convictions on his record, and I have the unmistakeable feeling that, once again, I have miraculously missed having a very dangerous confrontation.

It takes several weeks for me to begin to ease the incident from my mind. My safety and that of my family have been directly jeopardized. The first step I take is to install a genuine burglar alarm in the house. Then I drill and pin all the windows.

At this point I find myself with the nagging feeling that my actions are somehow not quite appropriate. I am clearly reacting in fear to recent events and I am about to become a prisoner in my own house. But it is difficult not to continually imagine disasterous futures — coming home with my little daughter and having to confront an intruder, then to kill or injure him or to have him kill or injure me. Or worse, my daughter. Or for her to have to even witness such an event. But thinking fearful thoughts, feeling afraid and locking myself in the house is not the solution. Once again remembering that I am what I think about all day long, a better solution is to find a way to release these disturbing thoughts from my mind.

There is a school of knowledge that elaborates on my journeyman-friend's observations about thinking. Proponents contend that the only way I can ever rid myself of the fears I carry around with me daily, is to first recognize my thoughts about bad people, and the things they have done or might do, as my thoughts. Such thoughts are separate from any actions that take place in the real world. Advocates call such thoughts "attack thoughts". Whether they concern attacking others, defending or being attacked, they are all considered attack thoughts. Being unaware that such thoughts are directing and coloring my perception leads me to react to circumstances and conditions that do not exist. I start considering putting bars on the windows. It is difficult to see that I am reacting to thought images that I have made up, and not really reacting to reality, particularly when my next door neighbor's house has been burglarized twice and mine has just had a burglar caught in it. But that experience happened moment by moment long ago. It's over and is being kept alive only in my memory as "attack thoughts".

Ridding myself of such thoughts is not easy. A slow, step by step process is required. The first step is to replace them with a saying I repeat to myself as often as I can remember — I can escape from the world I see by giving up attack thoughts. It's a welcome beginning.[3]

Though not of this particular school, Sheldon Kopp, a well-known, east coast, psychologist, asks this question in a similar vein: "How can we tell the difference between simply *feeling* scared and actually being in danger, or between only *feeling* secure and really being safe?"

Kopp describes the particular fear I am experiencing as "fear without":

"We fear sudden nuclear destruction, incurable disease, freak accidents, and random crime. We fear failure, loneliness, and death. Fear is the one emotion that traps us all, preventing the possible happiness that life has to offer. Running away may relieve our anxieties momentarily, but lasting ease requires our turning toward what we dread most. In dealing with fear, the way out is in."[4]

By turning in, we can learn to recognize real risk and danger in personal terms and evaluate for ourselves whether our fear is real or imagined danger, and justifies what we must do to manage. I must decide the nature of a danger for myself and whether the rewards for taking or avoiding a particular action are worth the risk to me. Whatever I do, it is a mistake to act out of a fear that my actions might make someone else uneasy or think I am weak or foolish.

When I think about for whom I am installing the burglar alarm and pinning the windows, I recognize a variety of people: I'm pinning them for my wife (who has directly expressed her apprehensions) and I'm pinning them for the child I once was (who is unable to handle the prospect of having to deal an intruder). I'm not really acting on behalf of my daughter, however. If any one of us can deal effectively with an intruder, I think she can — with both arms open, with all the grace, innocence and elegance of the following passage:

"During his final incarnation, the Buddha seated himself under the Tree of Enlightenment. There he practiced the meditation that would take him to the threshold of release from fear and other forms of suffering. When the Evil One assaulted him with terrible armies of dark and devouring monsters, the Buddha sustained his safety with a sacred movement, raising his right hand against fear. He then extended his other hand in compassion, and announced his decision to go on facing his fears in this life so that the protection of his understanding would remain available until the last of us had also attained enlightenment."[5]

It is with a small sense of enlightenment that I discard the metal pins, open the windows wide, and breathe in the early morning air. The feeling is exhilarating — as if the long hours spent getting a bit of interior trim to fit just right has resulted in extraordinary success.

17

"One of the holiest experiences is to get
outside your head for a few seconds
and realize the sun doesn't rise
and set in your armpit."
— Stephen of the Farm

Making Light the Work

One of the holiest features of New House is that there is plenty of light inside. There are large windows in every room and two foot by four foot skylights in the hall ascend straight up through both floors. Most modern homebuyers like lots of light in their homes.

Feeling adventurous on one Experimental Friday, I decide I will spend the night at New House camped out in a sleeping bag. The electric has been completely finished, the water works and there's a Port-A-Potty out in the backyard — like an old-time outhouse. When I get there I receive several revelations. New House, at night, has an enormously different feel than it does in the daytime. It is almost like I'm in a completely different house. Light from a porch lamp across the street pours in through the den and master bedroom windows. Together with New House's exterior lights, the glare is very disturbing. It turns out the kitchen and the dining room are overlighted and, the living room and den don't have enough existing light in them. Without daylight streaming in from the skylights, the halls are very dark. Because this house is so different after dark and, when I consider how much more time is spent in a house at night, I wonder why people even bother house-hunting during the daytime.

As I'm considering this, another revelation I have is that of the two dozen plus houses I've built, never have I spent time in a single one of them at night. It's like I've discovered a whole dimension to my work that existed before that I had no inkling about. What other secrets might be revealed, I wonder, if I stick with this work for the rest of my life? It's something to ponder and anticipate.

Given people's love of light, it is interesting to consider the assertion of today's physicists that all matter, living and dead, is made up of the same "stuff" as light:

> "The subatomic units of matter are very abstract entities which have a dual aspect. Depending on how we look at them, they appear sometimes as particles, sometimes as waves; and this dual nature is also exhibited by light which can take the form of electromagnetic waves or of particles."[1]

Houses, humans and sunshine are essentially made up of different forms of the same stuff! I take this as proof that, as frequently as we are able, we must make this hard, dense, heavy, work as light as possible. One way we keep our work light is through the telling and playing of good-natured, safe, practical jokes. Several examples:

When the mechanical contractors are soldering gutters for New House, I have them solder a long roofing nail to the back of a half dollar. Just before lunch I take the half dollar and hammer it into the wooden subfloor at the top of the stairs and then sit down to eat. Workman after workman walks up the stairs, comes to the top, looks at the half dollar, looks up at the group staring at him, and then simply steps over the coin and sits down to eat. It looks like the joke is on me. Finally though, just as lunch is almost over, along comes an unsuspecting realtor, wanting to know who owns the building, who is building it, is it for sale, etc., etc. He gets to the top of the stairs, spots the half dollar, takes a sidelong glance and then gives it a little kick with the side of his foot. When it doesn't move, he kicks it again, harder this time. When it still doesn't move, he bends down and tries to pick it up. When he can't, he looks over at us and finds a half dozen guys staring back with big wide grins.

We also like to tell stories. A friend of mine took and passed his contractor's licensing examination. The very first job he bid successfully was to renovate an old church. Initially, the job went along fairly well but, towards the end, it became apparent that he had seriously underestimated certain portions of the work. When all the renovation was completed and the church was ready for painting, my friend had just about run out of money.

To stretch the remaining money as far as he could, he went out and bought a single forty gallon drum of paint, thinned it down as much as he could, and began to apply it to the church. He thinned it and thinned it until there was very little actual pigment left in the paint. Then, just when the last bit of painting was done, it started to rain. Within minutes the rain washed all of the watered-down paint completely off the church.

My friend stood in the rain looking at the mess wondering what to do. Suddenly the sky thundered, lightning flashed and a great voice came booming down. It said simply: "Repaint . . . and thin no more!"

On another occasion Stan and I are in a lumber yard shopping for plumbing fixtures. On the way out of the plumbing department, we pass by various rows of paneling on display. One of the panels has a surface made out of simulated brick. Upon seeing this, Stan suddenly breaks out into a wild fit of laughter.

"Look at this brick paneling," he says between laughs. "Imagine what would happen if Sparky (a partner with Stan in another venture) was to come home tonight, open his front door and find that the whole thing has been bricked in! Imagine the look on his face."

Since Sparky has a face that is very good with funny looks it is not too hard to imagine. Stan is still laughing though, long after we pay for the plumbing supplies and are on the road back to New House. During the trip it occurs to me that, if this is such a funny joke to play on Sparky, imagine how much funnier it would be to play on Stan. Immediately I call up Sparky who shares an office with Stan. I find out what time Stan gets in to work in the morning and that Sparky is enthused about playing this little joke on him.

Several days later I pick up a piece of that same brick paneling. When Stan comes to work the following morning the panel is securely in place and we are all safely hidden in an office across the hall. We're ready with cameras and flash to record the moment as a cherished piece of trickster history. Stan opens the door and immediately knows he has been hoodwinked: the joke is on him.

A great source of housebuilding humor several years ago was Dave Barry's column, "Off Center", in New Shelter magazine. Here's a piece that I surely related to on my way to becoming a builder:

How to Make a Board
by Dave Barry

Most of what I know about carpentry, which is almost nothing, I learned in shop. I took shop during the Eisenhower administration, when boys took shop and girls took home economics — a code name for "cooking".

I regret today that I didn't take more shop in high school, because while I have never once used anything I know about the cosine and tangent, I have used my shop skills to make many useful objects for my home. For example, I recently made a board. I use my board in many ways. I stand on it when I have to get socks out of the dryer and water has been sitting in our basement around the dryer for a few days and has developed a pretty healthy layer of scum on top (plus heaven-only-knows-what new and predatory forms of life underneath).

I also use my board to squash spiders. (All spiders are deadly killers. Don't believe any of the stuff you read in National Geographic.) Generally, after I squash a spider, I leave the board in the water for a few days, spider-side down, to wash it off, assuming the scum isn't too bad.

If you'd like to make a board, you'll need:

Materials: A board, paint.

Tools: A chisel, a handgun.

Get your board at a lumberyard, but be prepared. Lumberyards reek of lunacy. They use a system of measurement that dates back to Colonial times, when people had brains the size of M&Ms. When they tell you a board is "two-by-four," they mean it is *not* two inches by four inches. Likewise, a "one-by-six" is *not* one inch by six inches. So if you know what size board you want, tell the lumberperson you want some other size. If you don't know what size you want, tell him it's for squashing spiders. He'll know what you need.

You should paint your board so people will know it's a home carpentry project, as opposed to a mere board. I suggest you use a darkish color, something along the lines of spider guts. Use your chisel to open the paint can. Have your gun ready in case a shop teacher is lurking nearby.

Once you've finished your board, you can move on to a more advanced project, such as a harpsichord. And when you get the hang of using your tools, you can make all kinds of other projects. Here are some I've made:

A length of rope.

Wood with nails in it.

Sawdust.

If you'd like plans for any of these projects, just drop some money in an envelope and send it to me and I'll keep it.[2]

Many of the subcontractors who come to work on my houses become good-natured victims of all sorts of mischief. On any day the plumbers or mechanical contractors might show up and find every person on the job walking around with a flat carpenter's pencil stuck horizontally in their teeth, or perhaps all wearing silly hats or Groucho Marx noses and glasses. We'll even stage a prank occasionally. For example, we once placed a plastic garbage can filled with foam boulders on the edge of a second floor landing, and then got the flooring contractor to agree to work below directly underneath it. Again, another unsuspecting realtor was on the job when we "accidently" tipped over the can. With the realtor looking on, I screamed, "look out", but it was too late. The way the floor guy collapsed with boulders all on top of him, the realtor thought he was dead for sure. At the time of an accident like that, the brain doesn't stop to think: "Now, why did those guys have a barrel of boulders up on the second floor, and how come they didn't make any noise and they bounced when they landed?" In this case, and in most cases, we know the person for whom we're staging the prank and are relatively sure they can take a joke.

I also have a prank chest in my truck filled with fake throw-up, severed arms and fingers, exploding pens, fake feces, fart spray and an assortment of hats with bizarre and idiotic sayings on them.*

Lunchtime at New House is also the favorite time for joke telling. One of my personal favorites is this:

One day Jesus is making his customary rounds in heaven when he notices a wizened, white-haired old man sitting in The Carpenter's Corner looking most unhappy. The next week he comes across the man again, looking equally miserable, and then finally, a week later he stops to talk.

"See here, my good man," says Jesus kindly, "this is heaven. The sun is shining, there's plenty of food, pneumatic tools, all kinds of wood — you're supposed to be ecstatically happy! What's wrong?"

"Well," says the old man, "you see, when I was a carpenter on earth, I lost my only, dearly beloved son at an early age. And here in heaven I was hoping more than anything else that I would find him."

Tears immediately fill Jesus' eyes. "Father!" he cries.

---

* *Two frequent sources of supply for my mischief are mail order catalogue houses: Funny Side Up, 425 Stump Road, North Wales, PA 19454, and the Johnson Smith Company, P.O. Box 255500, Bradenton, FL 34206-5500.*

The old man jumps to his feet, bursts into tears too, and sobs, "Pinocchio, my boy!"[3]

When the plywood siding is initially installed on New House, all except the largest doors and windows are covered over completely and only cut out later to the exact window dimension for when the windows arrive. This approach helps keep the building secure and makes for the fastest installation of windows. Until the windows are cut out, the house tends to be extremely dark and work within its walls is often done with the aid of interior lights. The change is quite dramatic when the windows are finally cut out and all the light comes streaming inside, which calls a story to mind:

During a routine check of his jobsite, a foreman finds an apprentice carpenter hanging from the exposed beams of a vaulted ceiling. "What the hell are you doing up there?" he calls out.

"Can't you see, I'm a chandelier," the apprentice answers.

"Well, get down and get back to work right now," replies the foreman. His orders are to no avail for, on subsequent checks, he finds the apprentice in exactly the same position. Finally, he has no choice but to fire him. That evening the foreman discovers all the other apprentices on the job are quitting — packing up and getting ready to leave for good.

"What the hell do you clowns think you're doing?" the foreman shouts.

"Hey, listen," they answer, "we ain't working without no lights."

Throughout the process of building New House, the crew and I like to demonstrate our prowess to each other with the dark arts, magic and illusion. Since many of the guys have the skills of master magicians when it comes to making one or another of my tools disappear, it feels like just desserts when I disappear someone's lunch and pull a never-ending paper chain out of their lunchbox or, even better, when I transform their hard-earned greenback dollars into a worthless bag of confetti.*

Timing is everything with a good trick or joke, and occasionally someone will pull something inappropriate or in bad taste, and from time to time something gets taken the wrong way. Usually the worst that happens is someone gets their feelings hurt. Eventually they get over it. When tempers do flare from time to time on the job, nothing relieves the tension like a "silly string war", where two or more combatants are each given a can of silly string and have at it. In my experience, its far better to err on the side of fun and play than simply have day after day of noses-to-the-grindstone labor. More often than not, if it's not light, it won't be right.

** Most of the magic we do is very elementary and done with inexpensive or even homemade props. Two sources of such items are Louis Tannen's Magic Supply, 6 West 32nd Street, NY, NY 10001, and Hank Lee's Magic Factory, 125 Lincoln Street, Boston, MA 02135.*

18

"When work is soulless,
life stifles and dies."
— Albert Camus

Waste Not, Want Not

New House is served by city sewers. A 12 inch diameter pipe, about seven feet in the ground, is buried in the middle of the street and connects to a 24" sewer main (to which all the other smaller sewer pipes are then connected). These pipes carry the waste out of town to a treatment facility.

Connecting up the sewer represents a major investment of time, manpower and money. Normally, the best idea is to get the sewer connected in conjunction the installation of the finished plumbing fixtures — toilets, tubs and sinks — along with all the other utilities. This way the builder doesn't have a large investment of time, money and materials tied up for six months or longer performing no useful function.

Because New House sits back from the road, I did a lot of investigative work to determine that the sewer could be connected with a minimum of headache and heartache. For New House, the sewer pipe can be black plastic if I have the required fall per foot of run. This requirement is roughly a ¼ inch drop per foot of pipe needed. From the stubbed out pipe in our foundation that means I need a total of at least four feet of fall. (*See Illustration 18.1*) So the sewer main must be buried at least four feet in the street, provided the land is level, which it is. A check with the sewer district determines that their pipe actually is buried seven feet in the street. Now is that to the top of the pipe or to the bottom? The engineer tells me its to the top, but even if it's to the bottom, I still have two feet of room for error. With this information, I feel confident that connecting the sewer will not be a problem, and I can proceed with the actual construction of New House.

When the time is right, a permit is still needed to connect up to the sewer system. I complete and turn in the application and the clerk then asks for a check for almost $4000!

"Well," I rationalize, "There's a lot of digging, and it can be pretty dangerous. Plus I bet their insurance for working in the road is expensive."

I am stunned when I call to tell the district people that I am ready to have them come out and hook up my sewer and they tell me that isn't their job. The $4000 is simply a permit and maintenance fee!

Illustration 18.1

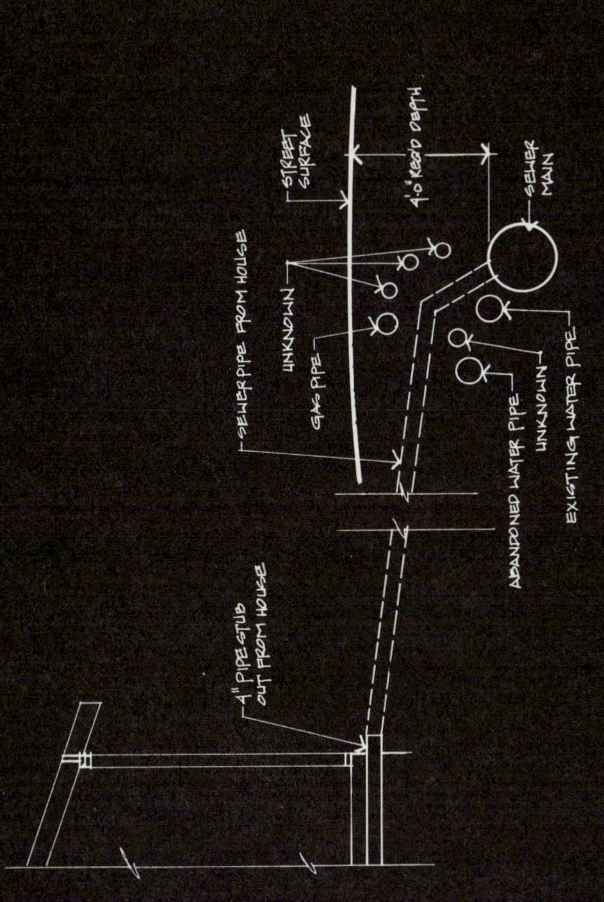
STREET SURFACE
4'-0 REQ'D DEPTH
SEWER MAIN
SEWER PIPE FROM HOUSE
UNKNOWN
GAS PIPE
ABANDONED WATER PIPE
UNKNOWN
EXISTING WATER PIPE
4" PIPE STUB OUT FROM HOUSE
REQUIRED SEWER PIPE ROUTING DETAIL

Since I am now responsible for digging and connecting the sewer, my first thought is to hire a backhoe, excavate and hook up to the main myself. The actual physical tapping into the main is usually performed by a person who holds a patent on the process. After a brief discussion with this gentleman, I am advised to subcontract out the complete installation to an experienced underground contractor. In street work many things can go wrong and often do. Among the things I must be sure to determine is that my contractor has installed sewers before and that he has adequate "tumbling" insurance (insurance that will protect him in case some passing motorist, driving by deep in trance, ignores his waving flagmen, and tumbles head-first into the open trench).

I call several contractors who come out to look at the job and never hear from any of them again. Finally, I remember a contractor that I once worked with many years ago. He comes out, looks at the job, and agrees to do the work . . . at a price more than twice what I originally budgeted. But waste not, waste not.

The night before we are to open the street I sleep fitfully. Images of exploding gas mains sending New House and the surrounding neighborhood up in flames fill my head, only to give way to images of those flames being quenched by the ensuing flood from the newly broken water main. Finally morning comes and Jake, the contractor, shows up right on time with his two sons. The sons will be doing the actual work.

And my growth as a housebuilder will be continuing.

In the concern and confusion about getting the sewer

put in at all, Jake never gives me a contract and I never ask him for one. We've worked together before and I know what we've agreed will happen: he's to put in the sewer and I'm to pay him $5500. I note that much in my daily log.

The first plan of attack is to dig a test pit in the street down to the sewer main and then dig a second pit back from the road and attempt to bore horizontally over to the main. This course of action must soon be abandoned however, when some other pipe is found to be in the way and blocking the direct line in. Consequently the whole street must be opened up and a trench dug back beyond the property line for the lateral. That is accomplished rather quickly.

Once the street is open, we are confronted with a total of eight pipes. The sewer pipe can be readily identified because of its size and how it lines up with a manhole cover further down the street. Unfortunately, it is only five feet to the top of the sewer. That still provides enough fall. We cannot identify the four other pipes — especially the 18″ diameter pipe that is directly in line with, and slightly above, the sewer main and completely blocks us from coming in at the prescribed point of entry.

Men from all the different utilities come out to identify the various pipes. Four pipes remain that no one can identify, but they are of no concern to us. However, the pipe blocking us is of concern. It turns out to be a feeder main for the city water supply, which is why it's so large. And my sewer line must go over and clear it by a minimum of six inches, meaning I will be losing almost two feet of fall! There is no creative alternative that makes economic sense: the lateral must go in where it must go in.

Once the lateral is actually installed, Jake's sons measure the elevations from New House's waste main and the lateral and tell me that we will be able to just make the required 2% fall. We all breathe a sigh of relief. Except that when they trench and lay in the sewer pipe with their laser level, the line ends up a foot and a half above my stubbed out house pipe!

At this point I feel myself starting to panic. What if I can't get a sewer connected at all? What if I've built a house in which no one will be able to live - a very large, very stately, and very expensive storage shed? My first call is to Inspector Dan to make sure there are no local ordinances that prohibit the installation of a sewage pump if that turns out to be the only viable solution. There are none. Then I ask him about the possibility of installing the line with a 1% fall. Dan indicates that 1% will be alright, only it has to be in cast iron.

Jake's sons indicate that they have discovered a problem with their laser level. They have corrected it and a 1% line can be laid with room to spare. I give them the go-ahead. Three hours later the trench is once again up to the house stubout. Only *six* inches higher this time. We're making progress, only we still can't connect. I find myself losing patience. It is clear that Jake's sons do not know how to determine elevations properly. After much hemming, hawing and confused conversation, I tell them to pack up their stuff and I'll speak to their father in the morning.

The next morning I go over the particulars of the work with Jake. He and I both agree there were problems and mistakes on the job and that the only solution at this point is the installation of a sewage pump because it doesn't need any fall in the line for its operation. He also tells me that, because of the delays in the street, he has some overtime that he'd like me to pay. I reluctantly agree. Three days later I get his bill for an additional $5500! He's billed me for the extra street time, as well as for the trenching and retrenching. Overall, he has doubled the original agreed-upon price. He must have gone to the same school to learn billing practices as our surveyor.

Now I'm really upset. I call Jake on the phone and he says his boys were there, the equipment was there, and he can't afford to lose money with all that heavy equipment and all those union wages. This is totally unacceptable to me. I tell him that I'd like a complete list of the men, materials and machines that he used on the job before I will agree to send him any money.

Weeks pass and I do not receive the written documentation for the work that Jake agreed to send. My fantasy is that he has reviewed his time logs and come to the realization that the $5500 he charged per the original contract has more than covered the extra time he put in on the job (which it has, by my calculations). I assume he has decided to admit that his boys made a mistake and that it would be unfair to expect me to pay for it. That's when I get his call.

Hearing his voice on my answering machine brings up a whole host of negative feelings. There is disappointment that my fantasy is not going to be realized. I feel aggravated that I have to continue to deal with this disagreeable situation. There's an undercurrent of anger and threat in his voice on my machine and I can feel my own anger rising in response. I listen to the message and decide to put off returning his call until the next day.

I ring up the painting contractor instead to come and finish a number of remaining areas that need to be touched up at New House. The ringing stops and a voice answers:

"Jake's Underground Construction." I have not dialed the painter's number, but have unwittingly dialed Jake's office! I am shocked awake and in the moment that it takes me to respond, I marvel at the workings of my own unconscious: a part of me must really want to get this thing resolved.

"Jake. Mark Brady returning your call."

"Yes, Mark. Listen I'd like to get this bill taken care of."

"Well, I thought we left it that you were going to send me a complete list of all the time spent and work done and we would take it from there," I answer.

"But the bill I sent you for the extras has all the time on it."

"Well, Jake," I reply, "here's the problem. Once the lateral was in and we knew what that elevation was, your boys measured the lateral and the house stubout with the laser level and told me they could trench and hook up to my stubbed out pipe and maintain the proper fall. They didn't say, 'We're not sure' or 'We may need to get someone out here to confirm that the line can go in.' or indicate in any way that there was a problem. We'd overcome the difficulties with all the pipes in the street and I had absolutely no cause for concern. From the lateral they then proceeded to trench and lay in the plastic pipe at 2% grade. When they made the turn toward the house and ran the pipe about half way, they measured the elevations again, only to discover that we were well over a foot higher than the house stubout. Many conversations took place. Then they told me they could get the pipe in at a 1% grade, pulled the plastic pipe out, and proceeded to re-excavate. Well, the 1% line ended up higher than my stubout too. At this point I sent your boys home, put the plastic pipe back into the trench and set it up for a pump system myself."

"First we trenched and *then* found out that the sewer main was blocked by all the pipes," Jake says. "That's why we had to retrench to a 1% grade."

"Jake, you've got the sequence all wrong. The lateral went in first and then the trench. Why don't you talk it over again with your sons and get back to me. I'm sure we can get this thing ironed out."

Throughout this conversation I notice my energy level rise and fall. As Jake relates what I consider to be factual errors in what took place, I find myself getting very excited and have a difficult time allowing him to finish what he wishes to say.

My ultimate aim here is to get this situation resolved in a manner that does not create separation and negative feelings. I'd like to work for some kind of a principled settlement. Roger Fisher and Scott Brown suggest I think of myself as a partner of Jake's, (or a brother-in-law?) and that we are in a hardheaded, side by side search for a fair agreement advantageous to each.

At this point my sense is that all Jake would like to do is have me pay him all the money he has billed me. He feels he's right and entitled to it. I feel I'm right and shouldn't have to pay it. I'd like this dispute to have a different feel and flavor than others in which I've recently engaged. While I'm not sure how to go about achieving that end result, I continue to hold the desire in the forefront of my thinking.

In the past, my usual way of operating in this situation would be to find some rationale for paying him the money and to forget about it. I'd want to be nice.

"I'm doing well and I can afford it, so what the hell."

"I expected to have to pay more, so I'm not really losing anything. Besides, I've saved money on other work."

"Well, I'll give the money with a charitable heart. Afterall, the more you give, the more you get."

Somehow these ways of abandoning the process never feel quite right. I feel mistreated, like I've discounted myself. Fisher and Brown call this "An Extreme Application of the Golden Rule", and consider such thinking a mistake. They list six components of my position along with the rationale.

- Rationality. Since I would like you to base your actions on love for me, I will base all my actions, not on reason, but on love for you.
- Understanding. Since I would like you to accept my understanding of the situation as correct, I will accept yours.
- Communication. Since I would like you not to bother me with problems, there is no need to talk about any of our differences.
- Reliability. Since I would like you to trust me completely, I will trust you completely.

● Coercion / Persuasion. Since I would like you to yield to me, I will yield to you.

● Acceptance. Since I would like you to accept my interest and views as controlling, I will accept yours as controlling.[1]

With this information as a guide, I resolve not to abandon the process. I am able to separate my feelings about Jake and his sons from the problem of being overcharged. Basically, I like all of them, and the mistake they made is one I continually caution myself and my help against: thinking and acting like I know something that I really don't know. After they failed to connect up the pipelines the second time, it became obvious that Jake's boys didn't know how to shoot elevations with a laser level.

As positions go, mine could easily be: "I'm right and you're wrong." I recognize a large part of me does think this. I also know such thinking can be a detriment, that it can seriously block my ability to focus on mutual interests. Fisher and Brown call the approach that results from this kind of thinking "An Eye For an Eye". The components and rationale for this position are:

● Rationality. Since anger dominates your thinking, it will dominate mine.

● Understanding. Since you misunderstand me, I will put the worst interpretation on what you do — a prescription certain to produce misunderstanding.

● Communication. Since you are not listening to me, I will not listen to you.

● Reliability. Since you are apparently trying to deceive me, I will try to deceive you.

- Noncoercive modes of influence. Since you are trying to coerce me, I will try to coerce you.
- Acceptance. Since you are denigrating me, my views, and my interests, I will denigrate you and yours.[2]

I do want to pay a fair price for the work done. The problems in the street with the conglomeration of pipes doesn't belong to Jake, so I should pay something extra for his time out there. It's also in my interest to resolve this amicably and have him available to do other work (I will obviously have his laser work double-checked in the future). In addition, the heavy equipment community is small in this area, and because of the trouble I had getting someone to put the sewer line in, it probably would not serve me well to have those people speaking badly of me.

Finally, it is in my interest to work this out fairly because it genuinely allows me to acquire and demonstrate new learning, and helps me to positively define who I am. With a successful resolution, I'll feel good and like myself just that much better.

Fisher and Brown suggest caution here. The position they recommend is called the "Unconditionally Constructive Strategy". The guidelines they offer are not to be considered advice on how to be "good", but rather, how to be effective.

- Do only those things that are good for the relationship and good for us, whether or not they reciprocate.
- Rationality. Even if they are acting emotionally, balance emotions with reason.

- Understanding. Even if they misunderstand us, try to understand them.
- Communication. Even if they are not listening, consult them before deciding on matters that affect them.
- Reliability. Even if they are trying to deceive us, neither trust them nor deceive them; be reliable.
- Noncoercive modes of influence. Even if they are trying to coerce us, neither yield to that coercion nor try to coerce them; be open to persuasion and try to persuade them.
- Acceptance. Even if they reject us and our concerns as unworthy of their consideration, accept them as worthy of our consideration, care about them, and be open to learning from them.[3]

To generate a variety of unconditionally constructive possibilites before deciding how to resolve our differences, I decide to send Jake a letter. First I outline the sequence of events as I remember them. I also send along a complete list of the hours worked and what was accomplished. I make no recommendations or judgements. In terms of money, I have no absolute clarity about what is correct. My sense is the amount we originally agreed upon verbally is probably appropriate. Little extra work would have been necessary if no mistakes were made. As one option, I suggest I would be willing to pay a somewhat higher rate for work that I could budget for in the future to help cover any deficits this job cost him. I also intimate that I'll keep my eyes open for other work opportunities that friends might have that would be right for him. I ask him to call me in a week.

Several weeks go by with no word from Jake. I interpret this lack of contact as him not knowing how to respond: afterall, his boys did make a mistake.

When I think about whether this conflict can be constructively resolved, it seems within the realm of possibility. We will both offer our points of view, hear each other out and explore as many options as we can come up with. One tack, Jake could take that would be most effective with me, would be to admit they made a mistake and ask me to pay some reasonable portion of the extras. I would probably agree readily and feel good about the resolution.

But best intentions and strategic resolutions don't always come to fruition in reality. Two weeks later I finally get a call. Jake's oldest son tells me that Jake has died ten days ago from an aneurism in his brain! He'd like me to please send them the money outstanding on their last invoice. So much for the best laid plans.

I arbitrarily subtract 25% from Jake's invoice and send his sons a check for the balance. Because it's the nature of who I am, I also decide to read the newspapers and check the local obituaries for the past few weeks and . . . Jake has indeed passed away.

But the intention and desire of my research and rehearsals remain. This time fate intervened and disappointingly cut the process short. During this period of grief and loss, it did not feel appropriate to carry on negotiations with Jake's sons. But other opportunities to consciously resolve other conflicts in an unconditionally constructive manner will undoubtedly continue to present themselves.

19

"Every building is given its character
by certain patterns of events that
keep on happening there."
— Christopher Alexander

Punching Out

As the building of New House winds to a close, one of the patterns that keeps on happening there is list-making. I'm continually making lists of all the little details that must be attended to before the project can be considered complete. These lists go by many names, "Things to Do", "Completion Check List", "Wrap-Up Items", etc. The name that is most appealing is "The Punch List". The name, "punch list" suggests that the items remaining are minor repairs or omissions that can be dealt with quickly and crossed off the list. Bam, bam, bam — just the way a mechanical press punches out washers or silver dollars.

In reality punch lists seem to serve only one purpose: to generate other punch lists, seemingly ad infinitum. A frustrating, repeating pattern, but one that teaches patience and builds character. Fritjof Capra might call the work of housebuilding and punch lists "entropic work" and describe it this way:

> "Doing work that has to be done over and over again helps us recognize the natural cycles of growth and decay, of birth and death, and thus become aware of the dynamic order of the universe. 'Ordinary work', as the root meaning of the term indicates, is work that is in harmony with the order we perceive in the natural environment . . . what we need therefore, is to revise the concept and practice of work in such a way that it becomes meaningful and fulfilling for the individual worker, useful for society, and part of the harmonious order of the ecosystem. To reorganize and practice our work in this way will allow us to recapture its spiritual essence."[1]

As I repeat all the tasks, the ordinary work that goes into the building of a new house, I occasionally sense that essence. The repetitive work that installing roof shingles requires, for example, often brings with it such a sense. It is a powerful experience, so much so that I thought to capture and contain it by taking an exam and becoming licensed as a roofing contractor. This was not quite a proper approach. As I have become more and more "licensed" and "professional" the experience seems to occur less and less in the present-moment flow of the work itself. More often I am able to recapture that essense in a relaxed, reflective moment, long after a project is complete. On such occasions I am reminded of the words of one teacher who pointed out how in all work — "Process is our most important product".

Making up punch lists and using them to work from is critical to the work of finishing New House. They organize and focus what would otherwise be an overwhelming amount of work. Often I begin a day doing one job, then I get distracted or interupted, and begin doing something else. Then other areas need my attention and soon several hours have gone by without me completing a single task. That's when I take out the punch list. I look it over, pick one task and mentally note that I will stay with it until it is totally complete. When that one is finished, I pick another and work it through to completion. The first tasks I pick are generally easiest, and this helps me get back to the work and get my focus restored. Focus and concentration are essential for me to feel connected to this work. Also, the more energy I am able to direct and channel to a specific task, the more energy I seem to have available to do that work. At the end of the day, I get a huge amount of satisfaction from being able to simply cross item after item off the list.

Whenever an item stays on the list for more than a week, it is usually a sign that its too large a job to be completed quickly. It usually requires that the task be broken down into a number of smaller jobs, or that some other work needs to happen first. For example, one item on New House's Punch List was: install island outlet. It's not a particularly difficult job, but it didn't get done for several weeks. When I took the time to think about that item, it became obvious that the materials needed were no longer on the job and needed to be acquired first. Once I purchased the materials the item could be promptly completed.

Because I most know what's needed, much of the work of punch lists, and much of my work as the general contractor on New House involves going to the store. Shopping. Discount houses hold the least amount of excitement and appeal for me. The crush of people and the merchandise in general disarray often mirrors my own internal disorganization. I prefer lumber yards, as they are more benign and benevolent on my psyche. The smells of the wood, the fir, pine, redwood and cedar, and the yards arranged in reasonably good order help rekindle my spirit much more than most discount shopping establishments.

But the large, local hardware store is the place that I almost always find rejuvenating to visit. I love to go there at 7:30 when they open, or around 8:30 just before they close at night. At those hours I encounter the fewest customers, cashiers and clutter.

On one particular Friday morning there's a certain peace and order that I experience the moment the electronic eye slides the door open for me to enter. Above my head are rows and rows of insulation batts stuffed into the ceiling. Together with the pipes and beams, they are all spray-painted a uniform primer red. As I walk by the line of cash registers at the front of the store I nod to several cashiers who know me by sight and feel the alternating heat and cold of the overhead heaters spaced evenly overhead. Although I have a number of specific items to purchase, I have deliberately given myself the experimental freedom to browse. Or more accurately, to experience the "tao of browsing": to simply empty my mind and cruise the aisles with open, non-critical observation.

For some reason browsing feels best begun at the left end of the store among the shovels, rakes and fertilizers. As I walk the aisles, my eyes are open for new experiences, tools, or materials I have never seen before. The first item I pick up is a new axe handle. When I acquire a large roll of fencing and two 9-pair shoe racks, I must replace the hand basket with a rolling carriage from the front of the store.

I return to the plumbing section and know that I will find Dave, the floor clerk, patiently stocking and organizing the shelves. We chat and, once again, he reiterates his disdain for the masses who descend daily and turn the three aisles for which he is responsible into plumbing stew every day. Each evening sends him home in despair at having failed to stem the raging tides of chaos.

As I move on to the electrical section, I make a mental bet with myself that I will not find anyone to assist me in any of the three electrical aisles. The first aisle is indeed empty. So is the second. In the third aisle the bet is lost when I spot the familiar blue shirt worn by all the store employees. Only it turns out to be a customer. He looks over at me and asks where he can find two-gang, outdoor GFI box covers. I assure him I don't know and then realize that I too am wearing the same blue shirt. I know there is supposed to be someone working in electrical at all times but, inevitably, when I visit that section, no one is ever there. But perhaps that customer is right. Perhaps it is supposed to be me. Holding that thought, I proceed to draw, measure, cut, price and tag my own wire. When I am done I realize: it is supposed to be me. I have both won and lost the bet in the space of a simple realization.

The next aisles — Housewares and Painting — hold items of little interest for me. These aisles contain items that are used mostly by painters or people already living in a house. I have never made a purchase from the Housewares and Painting aisles. Instead I perceive them only as standing in the way of Tools & Hardware. Those are the aisles that truly move me. I slowly amble up and down, a half step at a time. I pick up several of the hammers, then lift and smell the tool belts. The leather of each one smells like every baseball glove I ever owned as a kid. I come to the utility knives section and immediately think about my own design for these handy devices. Since I go through two or three of these flimsy, mass-produced tools a month, my design would be much sturdier. It would be made of stainless steel, with interlocking halves that would hold the blade securely. I feel sure there is a market for a sturdy utility knife, no matter what the price. As I'm lost in admiration for my own design, I spy a new model knife on the racks. It's my exact design, only its not made out of stainless steel. I put three in my carriage, and feel gratified that I was correct in assessing the need.

At the right end of the store in the final aisle is the Industrial Department. It's a wonderful area to explore. My fantasy is that it is here that Industrial Light and Magic (the movie people responsible for the special effects in Star Wars, Indiana Jones, etc.) come to buy all the components critical to the success of their work. I buy several critical to mine: a roll of rope to pull the telephone and cable TV wires through their underground conduit, a length of dishwasher hose, and a roll of plastic gutter shield. It is time to check out.

I estimate that 30% of the time I check out of one of these stores, the clerks mischarge me for the merchandise. When they overcharge me, if the money is significant, I will point it out. Five dollars or less, I won't mention it. It takes more than five dollars worth of my time to get the mistake corrected. But usually they undercharge me. I used to point the error out each time it would happen, but the time necessary to get the undercharge corrected soon became more time than I wanted to spend. So I made this decision — any time I'm undercharged for merchandise, I note it on the receipt. When I go through the month's bills, I tally the totals and make out a check for the amount of the undercharges and send it to a local charity. It saves me time and hassle, keeps the money in circulation, and goes to work for a good cause.

As I prepare to checkout on this day, a number of store employees are congregated at the customer service counter. They are close enough for me to read their white plastic name badges and to overhear their conversation.

Carla: "How'd it go with you and Doug last night?"

Giana, smiling: "Fine. We had fun."

Brian: "Fun? What kind of fun?"

Giana: "Private fun, which you'd like to know about."

Brian: "I thought I saw Doug's car and your car here in the parking lot when I drove by last night. It was after ten o'clock."

Carla and Giana give each other a secret look and then both laugh loudly.

Witnessing this interchange feels like visiting a small town. Here's a group of people who know each other, who meet and work together daily, who form relationships, and who do things together outside the store. Serving customers and selling merchandise is only secondarily what is happening. First and foremost, people are working things out on a wide range of levels. Doug and Giana are in an intimate relationship. Brian is excluded. There's a hierarchy too. Cashiers are at the bottom. Stock clerks are next in status, then the assistant managers, and finally the head manager. He's like the mayor of this small town. The assistant managers are the police force. I am the stranger passing through. It's all grist for the mill.

As the morning grows later, the store begins to fill with more and more people. The volume of noise around me increases and everyday awareness returns. I'm overloaded and distracted and my thoughts are back at New House, wondering how things are going. I charge my purchases and load them into my truck. It's time to leave town. Time to move on.

As the list of items required to finish New House gradually begins to diminish, I notice that Mitch, Ramon and Carlos begin to work more slowly, with less focus and attention to what needs to be done. This is a pattern that repeats itself in many of the houses that I've built. Discussions with other housebuilders fails to confirm similar experiences, so I suspect *I* might be the crucial variable here.

There is something difficult about completing a work. When building a house has been an important part of my daily life for close to a year, I find it difficult to finish up and let go. My mind and body yearn to linger, to remain with the familiar, the comfortable. I know intimately every house I've ever built — every design flaw, every success, every defect and every aesthetic victory. Each has its own sights and smells, its own textures and experiences. It is always disturbing to leave. It's much like leaving a relationship with a person I have lived and shared my life with over a long period of time. I know that the work is done and I must be going, even though I don't know what's next down the road.

Somehow, the guys on the job pick up my reluctance and do their best to bring the end as slowly and painlessly as possible. But, inevitably, the punch lists get done, every room is completed and there is simply no longer any logical reason to come to this particular place to work any more. I will still drive by from time to time, just to see how the old house is holding up and to admire an accomplishment about which I can feel some genuine pride. I can feel good about the fact that I have provided a place for people to live, to retire from their work at the end of the day to rest and recover.

On one of the last days, as we are picking up to leave, I find myself feeling irritated at a remark Mitch casually makes. In response to an obvious look of admiration as I stand back looking at New House, he says:

"What's so great about this house? It's not exactly low-income housing or anything?"

I have no answer for him at the time, except that I

know building New House at that moment in that place is the right thing for me to be doing. All I can tell him then is, "Hey, the rich have to live somewhere, too!"

As I reflect on it, I first have to admit that no, this is not low-income housing. Instead it is a house that will sell for three quarters of a million dollars. And that is something to feel good about. Earl Nightingale points out that based on his own observations, simply using level of affluence, you can walk down most any street in America and use it as a rough gauge to measure the level of service that those people are providing to humanity. The more affluent the neighborhood, the larger the companies and the greater the number of people to whom these neighbors are being responsible. The more support they are providing, the more support they are in turn being provided.[2] This notion directly conflicts with ideas I have held about wealthy people for many years. While there is obviously not a hundred percent correlation, in my own circle of well-to-do acquaintances, the percentage is surprisingly high.

Gary Zukav, a well-known physics writer, speaks about these same people in more ethereal terms:

> "When a person reaches for authentic power and chooses consciously to bring that power into levels of interaction that are shared with other people . . . as the frequency of their consciousness increases, as the quality of their consciousness reflects the clarity, humbleness, forgiveness and love of authentic power, it touches more and more around them . . . as they shine brighter, as their light and power increase with

each responsible choice, so does their world."[3]

People who exhibit a repeating pattern of reflecting clarity, humbleness, forgiveness, and the love of authentic power, those are the people I am building for. When the work is finally finished, when the punch lists are completed and we have all moved on, it is for such people that I expect New House will ultimately be home.

20

> "There is a certain Buddhist calm that comes from having money in the bank."
> — Ayn Rand

How's the Money?

The decision to construct my first house on speculation is deliberately designed to address different issues involving work and money in my life. From the time I was born until I was twenty-two, I had very little cash and a lot of ignorance about money. I lived on welfare until age eighteen and, even today, I still go to the mailbox the first of every month anticipating that brown envelope with the green check from the State. It hasn't come for almost 25 years, but habits associated with survival die hard.

As a teenager scarcity of money generated a host of negative social consequences, for example, having to wear the same clothes to school every day for weeks at a time. It wasn't until I was 21 that I began to realize a lack of money was something I could positively change. So, three years out of high school I made a decision to become productive and self-sufficient. I managed to accomplish this to a reasonable degree and, along the way, I've studied money's laws and effects. I have been deeply moved and inspired by a number of writers on the subject. Eric Butterworth is one:

> "As one person gets off welfare and begins even to gear his thinking toward productive work, there is a rippling effect through the entire economy of the nation, even of the world. True, the influence is small, certainly imperceptible, but it is real. It is much like the phenomenon described by astrophysicists: if one person waves his hand he sets off a rippling effect in the atmosphere that has an effect on the farthest star. So as one or two persons here and there across the land begin to *think* work, *think* productivity, and *think* abundance, something happens. People become more secure and begin to purchase things; business begins to expand and thus takes on more workers; government pays out less in welfare and takes in more taxes . . . leading to economic health for the nation, which in turn becomes an affluence for prosperity for every person."[1]

Further learnings from my studies suggest that money is best conceived and experienced as stored or potential energy. An aspect of money I call "flowpower" characterizes the uses for which I can trade my life energy and to which money can be put. The more money I have the more flowpower potential there is, and the more potential there is the more choices and options are available to me. Accumulated assets are like the energy stored in plutonium rods in a nuclear reactor. Given the proper structure and context, money provides the fuel for producing and channeling abundant energy. It is the energy of money that facilitates organizing the manpower, materials and motivation to build New House.

In terms of Maslow's "needs hierarchy" already mentioned, before I achieved any success, I would often get the horse before the cart. Or, as a builder, to get the roof on before the walls. I would attempt to move whole hog into developing matters of the heart and soul, before I had properly taken care of matters of the stomach and back, survival requirements like food, shelter and clothing. Fear and scarcity-based attempts at building and other projects have always lacked flowpower. Building New House at last, does not.

The amount of money moving through my checking account during the six months that New House is under construction averages in the neighborhood of $25,000 per month. It takes me a while to get used to having sufficient money on hand each month to pay all the bills and then some. Managing it properly certainly requires some sense of Ayn Rand's Buddhist calm. Michael Phillips, asserts there are seven laws surrounding money and, understanding them can help provide that calm. I've already discussed his first law. His Second Law is: "Money has its own rules . . . the first and clearest is you have to keep track of your money."[2]

A financial software program for my Macintosh computer is invaluable for working with this law. New House is broken up into over 100 financial categories, but only 54 have money allocated to them. (*See Illustration 20.1*) Each month I total the expenses and enter them into the respective numbered categories. At a glance I can tell how actual costs are approximating projected costs. Before framing is complete, but after the rough electrical and plumbing are finished, I know that total electrical and plumbing are going to go over budget by at least 25%. With that realization, Stan and I talk it over and decide to do as much of the work as possible with Mitch, Ramon and myself. In the end, as a result of being able to make early, informed decisions, we are only 10% over on plumbing, and we are within budget on electrical.

Money's effects are as vast and varied as personal imagination and experience will allow. They also provide powerful opportunities for continuing my education:

As we are nearing completion, Stan tells me he is divorcing his wife and selling the house they built together. He also indicates that he is interested in buying New House. My first response is relief. I will not have to sit around with the house on the market for weeks or months waiting for a buyer. I can turn it over to Stan — he can buy out my half interest in the project and I can move on to the next work. I am excited and eagerly look forward to completion of the transaction and receiving the fruits of my labor in the form of a large equity check. Except it's not quite that easy.

Stan's decision to purchase New House makes our relationship more complex. No longer are we simply Stanmark Development building a house for some anonymous Silicon Valley computer tycoon. To build for and sell to

Illustration 20.1

NO.	ITEM	AMOUNT	NO.	ITEM	AMOUNT
1.	Temporary Utilities		66.	Luminous Ceiling	
2.	Misc. Rentals		67.	Built-in Appliance	
3.	Guard Service		68.	Fire Extinguishers	
4.	Cleanup		69.	Elevator	
5.	Demolition		70.	Pool	
6.	Rough Grading		71.	Solar	
7.	Erosion Control		72.	Plumbing	
8.	Drainage		73.	Ventilation	
9.	Utility Trenching		74.	Heating	
10.	Sewer		75.	Air Conditioning	
11.	Water		76.	Fire Sprinklers	
12.	Curb/Gutter/Sidewalk		77.	Electrical	
13.	Asphalt Paving		78.	Fixtures	
14.	Fencing		79.	T.V. & Antenna	
15.	Finish Grade		MISCELLANEOUS		
16.	Landscaping		80.		
17.	Street Lights		81.		
18.	Street Signs		82.		
19.	Reinforcing Steel		83.		
20.	Foundation/Footings/Slab		84.		
21.	Walks & Driveways		85.		
22.	Lightweight Concrete		86.		
23.	Struc. Slabs/Columns/Beam		87.		
24.	Tilt-up Walls		88.		
25.	Block Walls		89.		
26.	Veneer		90.		
27.	Fireplace		91.		
28.	Rough Hardware		92.		
29.	Steel Columns		93.		
30.	Steel Stairs & Rails		94.		
31.	Metal Fireplace		95.		
32.	Metal Doors & Frame		SOFT COSTS		
33.	Metal Service Doors		96.		
34.	Panelized Roof Struc.		97.	Plans/Arch.	
35.	Glue Lam Beams		98.	Site Engineering	
36.	Wood Siding		99.	Soils Testing	
37.	Trusses		100.	Permits	
38.	Rough Lumber		101.	Fees & Assessments	
39.	Rough Carpentry		102.	Model Furnishings	
40.	Finish Carpentry		103.	Model Maintenance	
41.	Finish Lumber		104.	Model Utilities	
42.	Wood Doors		105.	Advertising	
43.	Cabinets		106.	Sales Expense	
44.	Waterproofing		107.	Contingencies	
45.	Roofing		108.	Overhead	
46.	Sheetmetal		109.	Supervision	
47.	Insulation		110.	Profit	
48.	Skylight		111.		
49.	Store Front/Sash/Glazing		112.		
50.	Windows & Sliding Doors		113.		
51.	Mirrors				
52.	Enclosures				
53.	Plaster/Stucco				
54.	Drywall				
55.	Painting				
56.	Formica				
57.	Simulated Marble				
58.	Ceramic Tile				
59.	Resilient Flooring				
60.	Wood Flooring				
61.	Carpeting				
62.	Suspended Ceiling				
63.	Garage Door		TOTAL CONSTRUCTION COST		
64.	Finish Hardware		I CERTIFY THAT THE FOREGOING IS COMPLETE AND		
65.	Bathroom Accessories		ACCURATE TO THE BEST OF MY KNOWLEDGE.		

Borrower	Date	Contractor	Date

such a person involves a relatively straightforward process: we build the house and include in it those standard and extra features that we think will produce the nicest house for the widest assortment of buyers at the highest price. We will purposely leave out those features and extras which, while they might be nice, will not produce sufficient return to warrant including in the house.

An example would be lighting fixtures. Lighting fixtures come in a wide range of styles and prices to suit a broad range of tastes and pocketbooks. No matter what type fixtures we choose, the chances are very high that, whoever ultimately buys the house will not like some of them, and could conceivably change them all, simply to put their own signature on the house. We could leave the fixtures out altogether and give the buyer a lighting allowance except the Building Code requires fixtures to be in place before a final inspection can take place. Consequently, the best solution is to obtain low-to-moderately priced fixtures and install them for Mr. Anonymous Buyer.

With the buyer no longer being anonymous, but a friend and associate, lighting fixtures are no longer a simple select-and-purchase decision. They become one measure of the worth, and a barometer of the strength and nature, of the friendship.

With our relationship changed from being equal partners building a new house on speculation, to mostly me now building a custom house to meet Stan's needs and requirements, each of our roles becomes more fuzzy and complex. How, exactly, do we decide what custom features the house will have? What work should Stan do as partner and what should he do as buyer? How do we accurately and fairly determine how much he should pay? As the buyer he will certainly deal with himself less firmly than he would if someone wanted to buy New House from the general community. How do I deal with the loss of an ally in the building and selling process?

The first step to avoid this complexity is to be clear with each other that the roles have changed. No longer is it Stanmark Development building a house for sale. It is now Stanmark building a house for sale and Stan buying it, thus requiring that Stan operate in two separate and distinct roles often in immediate succession. Not an easy division to make.

Given these concerns, I begin to have second thoughts about the difficulty of selling the house to Stan. Perhaps the clearest and fairest course would be to encourage him to find another house to buy, particularly since he expects only to be living in New House for two years at most. Finally, I elect instead the more difficult course, realizing there will be much to learn and new growth required. I will sell my half of New House to Stan.

Once we make this agreement the growth process heats up: real resentments and conflicts forge to the fore. Stans wants earthquake braces installed at no charge at the entry that are not indicated on the plans. He wants a washer and dryer included, along with a $3000 refrigerator, top of the line kitchen appliances, bath fixtures and custom closet systems in each bedroom. I hold that we wouldn't be installing these for Joe Buyer and that he should pay for them as extras. The relationship grows tense and the work is slow and strained. He cannot step out of his role as buyer and back into the role of partner/ builder, and I seem to only be able to sustain the role of adversary.

Many sleepless nights ensue.

I go to friends and books for sanctuary and respite. Along the way I remember: "I am never upset for the reason I think."[3] This suggests my problems with money are rarely actually problems about money. In my life, they have more often involve issues of self-worth, direction, energy, desire and unresolved childhood issues. So I am not surprised when, ultimately, I realize what I am most angry and resentful about is not the money, it is the withdrawal and loss of support. Stan and I began the project with a commitment to be partners and allies, to work together to complete the building of New House. Stan's decision to become the buyer and my agreement to become the seller left me feeling abandoned and alone.

It doesn't take a psychoanalytic genius to recognize that I am unwittingly reacting to the experience of an Abandoning Father — the same one who opted out and left me fatherless at age four. While it is not surprising to discover money is not the real issue, it is certainly a surprise to discover that our conflict is directly connected to my early childhood loss and abandonment.

Once this discovery is made, the completion and sale of New House begins to move forward a bit more easily. Stan and I each have an independent appraisal made of the house's value and agree to split the resulting small difference. We make up the closet systems on the job, get the refrigerator, split the cost of braces, downgrade the kitchen and bath fixtures and Stan buys a washer and dryer from his ex-wife. Two years later Stan sells New House used, for considerably more than he paid for it, but slightly under what the area housing market has appreciated overall.

21

"Focus your Self, Luke, and all other things
will come to you!"
— Yoda, Master of the Jedi

Receiving the Fruit

At last is the day for final inspection. All the plumbing fixtures are fixed, all the lights are lighted, and all the doors are doored. Already I have been waiting for three days to call for the inspection, and, each day we have had to postpone it because we lacked the wrought iron hand-rail required to keep people from inadvertently stepping off of the second floor and falling down to the first. Finally, the iron has arrived and at last the rails are railed.

Simultaneously, we have come to terms with the neighbors, Brenda, Larry and Homer. As the landscape work was finishing up, Homer came over one morning and told us he thought New House looked "quite remarkable".

He even went so far as to tell me he appreciated the fence we put in. Brenda and Larry had Stan and I over to their house several times so we could see the "problem" — the view of New House from their windows. It was extremely difficult to be sympathetic. What about the view of their house from our windows? Eventually, through a lot of listening and attempting to come up with options and the win-win solution that eluded me when Jake died, we all come to terms. Stan and I presented Brenda and Larry with several of the color options that we were considering for the outside of New House. They chose the same colors that were our first choices, and life suddenly became easier for all of us. I actually think Brenda and Larry got used to New House and maybe even learned a little bit about their own stuck places through the process.

We have scheduled a morning inspection and, at the same time, the movers will be at Stan's apartment loading up the remainder of his furniture and belongings, preparing to bring them over and move them in. By 11:30 AM though, no inspector has shown up and I am in a state of high tension. Stan has called three times to see if we've passed, he's called the building department twice, and I am meticulously going through the house room by room to see if there is anything that we may have missed.

Finally, at ten minutes to noon, Diane, the lone female city inspector arrives. She takes a quick walk through the whole house, checks to see that all our electrical circuits are identified and operating, that our appliances are vented and working, that the glass doors are on the fireplace and that the rail is secure and measures 36" from the floor. Then she takes out a yellow tag, signs it and sticks it on the gas pipe where the meter goes.

"The utility people will be out this afternoon to hook you up," she says. "Nice job." And with that, she gets in her car and drives off. The whole process takes less than ten minutes.

I feel strangely disappointed. It seems so anti-climactic. I would almost prefer her to stick around to dig, poke and prod and come up with a whole list of things wrong. Or at least tell me all the many things that are right. In one respect this whole project is designed and built to specifications required by the city. In part we're building this house for them. It's not rational, but I want their representative to spend more time looking around the place . . . with awe and appreciation. That would be nice.

"What a fantastic house, Mark. So many fine features. Look, a computer nook complete with a telephone jack and a dedicated, grounded outlet. And hey, a telephone jack in the bathroom. Now that's a great idea. You don't know how many times the phone has caught me in the john. The people who buy this house can take "power showers". Wow, nice use of space in that laundry room. Looked pretty cramped on the plan. Gee, that arched window in the master bedroom turned out well. Golly, Mark, thanks for doing such good work and making my job so easy. I can't wait until you start your next house!"

Now that would have been a response more commensurate with the time and effort I have invested. And it also would have been very wierd. I mean, why should she care about our house more than any other. She probably looks at sixty or seventy houses a week. If she started lauding every one of them, she'd quickly become known around town as "Wierd Di."

But still some genuine acknowledgement is missing . . . from me. To the guys. And to myself.

With Carlos and Ramon, I arrange to have each of them send flowers to their wives for their kindness and consideration, and I pick up the cost. I also give each of them a sizeable cash bonus and I explain to them that they've both done good work and were indispensible to the project. I tell them I look forward to starting the next house with them in the near future.

I Mitch tell the same thing, and I give him a much bigger bonus. Thousands of dollars, in fact, for we have decided to do the next project together on some property that Mitch has already acquired and, for which he is in the process of having the plans drawn up.

Before I can express my appreciation to Stan, I need to give myself some personal time for self-appreciation and acknowledgement. I am a different person at the completion of New House than I was at the beginning. The temptation is to say "better", however experience has shown me that such judgements are not useful. And that is one of the ways in which I am different: I am less prone toward negative judgements. Although positive judgements may hold the same peril, I am admittedly more prone toward them now. While my mind is, by no means, pure and judgement-free, I am less bogged down by the baggage that judgements carry. I am lighter.

I am also richer. Except for my last days in the manufacturing business, in the nine months it has taken to develop and give birth to New House and to sell it to Stan, I have made more money than at any other period in my life. I have a savings account once again. I own stocks. I pay taxes. I have discretionary income. I can absorb lulls in the contracting business. I can take a vacation. I can take six months off if I want. I can once again live the perilous, comfortable, middle-class life.

I have always placed great stock in the often quoted lines by Johann Wolfgang von Goethe:

"Whatever you can do, or dream you can, begin it.
Boldness has genius, power and magic in it."

While I thought and felt I could assume the necessary responsibility for getting New House completed, I don't remember ever feeling "bold". What I felt was a sense of self-confidence and trust. Confidence that no matter what trials and tribulations may arise (and they will), and no matter how unwilling or unprepared I may be (which will often be the case), things work out. Or more specifically, I facilitate their working out. From time to time I may not like or understand the working out, like the problems New House had with its neightbors and the daylight plane, but I can get problems resolved. With increased trust comes a greater willingness to suspend judgement without the need to have to immediately understand everything. I can, as they say, "hang out" more. I can experience some things as wonderous and perhaps a bit mysterious.

One of the most mysterious and challenging aspects of New House has been the role of Stan as one of my teachers. Only occasionally have I touched on small episodes in my educational association with Stan. Thanks to the wonders of psychological transference and counter-transference, each day Stan showed up on the job he would seem to appear in the guise of my mother, father, sister, or any of a number of other significant characters from my early life. In any of these guises he would unwittingly appear offering me new lessons and hard learnings. Like Carlos Casteneda's Don Juan, one day I might find him warm, helpful and caring; on another day, harsh, critical and demanding. Sometimes I'd find him funny, friendly and supportive and an hour later, like a chameleon, he might be cold, caustic and monomaniacal in his self-absorption. But however I would find him, I would eventually know that, at least in some degree, I was always seeing reflections of myself. It wasn't by accident that we looked so much alike.

The night before we are scheduled to have our final inspection, I have a telling dream. I am with Stan out at the Palo Alto Municipal Golf Course. Our aim is to have fun. Stan is wearing a golf hat that has the name Stanmark Development on it. He takes a shot and then, instead of me taking my turn, I find Stan is shooting again. He tells me sharply to pay attention. This time he's wearing a different hat, one that says Goodmark Development. The next time we both hit balls simultaneously. Mine somehow goes right through New House's open Master Bedroom window, bounces down the hall, through the back door, and rolls into hole which is in the back yard. Stan's ball goes completely over the house but comes down in the back yard, too. His ball also rolls into the hole. We've both made holes-in-ones. I am amazed at this incredible feat and, when I look up from the cup, Stan's hat this time says, "Hit the Mark Development".

On a planet-wide scale we have really only added one

more house to the global village. Of the millions of new homes that are built on the planet each year, the accomplishment has little significance. Except that for me at this time, in this place, it is the absolutely perfect work. I have hit the mark. I have found enormous growth and learning in working with Stan, in working with all my helpers, and in the process of building New House. I have also rediscovered and recovered much that has always been of value in myself.

When I meet Stan at New House late that July afternoon, a short passage from J. Krishnamurti's, *You Are the World*, echoes softly in my mind:

"In oneself lies the whole world, and if you know how to look and learn, then the door is there and the key is in your hand."[1]

As I stand in front of New House, fully completed, I realize that I will be here as the builder for the very last time. I give my key to Stan and repeat the last of those words aloud:

"The door is there and the key is in your hand."

At some level the full meaning is never lost.

Notes and Resources

Preface

1. While most of the book chronicles the people, places, and problems involving New House, the house and the people are, in fact, composites constructed from various real people and assorted real houses.

2. Shah, Idries *Learning How to Learn*. New York: Harper & Row, 1978, pg. 53.

First Works

1. Adapted from a story that appeared in *Chop Wood, Carry Water* by Rick Fields et al. Los Angeles: J. P. Tarcher, Inc. 1984, pp. 123-124.

Preplanning Preparations

1. Shah, Idries "The Increasing of Necessity", in *Tales of the Dervishes*. New York: Dutton, 1970, pp. 195-197.

2. Phillips, Michael *The Seven Laws of Money*. New York: Random House, 1974, pp. 1-25.

3. Fadiman, James *Be All That You Are*. Seattle, WA: Westlake Press, 1986, pg. 17.

4. Pearce, Joseph Chilton *The Magical Child*. New York: Bantam Books, 1977, pg. 114.

5. Murphy, Joseph *The Power of Your Subconscious Mind*. Englewood Cliffs, NJ: Prentice Hall, Inc., 1963, pp. 28, 54, 64, 96.

Drafting the Plans

1. Wheelis, Alan *The Illusionless Man*. New York: Harper & Row, 1966, pp. 3-34.

Laying the Foundation

1. Fritz, Robert *The Path of Least Resistance*. Walpole, NH: Stillpoint Publishing, 1984, pp. 135-136.

2. Butterworth, Eric *Spiritual Economics*. Unity Village, MO: Unity School of Christianity, 1983, pp. 78-79.

3. Ibid, pg. 80.

Under Critical Eyes

1. Phillips, Michael *The Seven Laws of Money*. New York: Random House, 1974, pp. 1-25.

2. Bolles, Richard *The Three Boxes of Life — And How to Get Out of Them*. Berkeley, CA: Ten Speed Press, 1981, pp. 37-69.

3. Ram Dass *The Only Dance There Is*. New York: Anchor Press, 1974, pg. 80.

Setting the Framework

1. Hughes, Kathleen *"A House Constructed in a Few Hours Isn't Home Sweet Home"*, Wall Street Journal, January 26, 1987, pg. 1.

Taking It Higher

1. Berne, Eric *Games People Play*. New York: Grove Press, 1964, pg. 85.

2. Goldstein, Joseph *The Experience of Insight*. Boulder, CO: Shambhala, 1983, pp. 12-13.

3. Ackoff, Russell *Management in Small Doses*. New York: Wiley & Sons, 1986, pg. 46.

Fleshing Out the Frame

1. Ram Dass *The Only Dance There Is*. Garden City, NJ: Doubleday, 1970, pp. 151-152.

Raising Up the Roof Beams

1. Brady, Mark "Freudian Slip" *Fine Homebuilding*, No.43. P.O. Box 355, Newtown, CT: Taunton Press, Jan, 1988, pg. 114.

2. Anonymous, *A Course in Miracles: Workbook for Students*. Tiburon, CA: Foundation for Inner Peace, 1975, pg. 277.

Building Warriors

1. Trungpa, Chogyam *Shambhala: the Sacred Path of the Warrior*. Boulder, CO: Shambhala, 1984, pp. 25-41.

2. Robbins, Anthony *Unlimited Power: the New Science of Personal Achievement*. New York:Simon & Schuster, 1986 pp. 320-331.

3. Bush, Mirabai *"Fourteen Lessons I've Learned Trying to Manage a Conscious Business"*. in Fields, et al, Chop Wood, Carry Water. Los Angeles: J.P. Tarcher, 1984, pg. 117.

4. Pyle, Leon *Sterling Homes Policy Manual*. Sterling Homes, 1655 So. Main Street, Walnut Creek, CA, 1989.

5. Fulghum, Robert *Everything I Needed to Know I Learned in Kindergarten. New York: Villard Books, 1988, pp. 6-7.*

The Heart of the House

1. Editorial, *"The Builder: Out of Chaos, Order."* Home Magazine, Volume XXXIV, No. 2, pg. 63.

Experimental Fridays

1. Schumacher, E. F. *Good Work*. New York: Harper & Row, 1979, pp. 3-4.

2. Stone, Don and Black, Les *"Right Livelihood: Doing It Right"*. New Age, Volume 4, No. 4, September, 1978, pp. 58-63.

3. Tart, Charles *Waking Up*. Boston: New Science Library, Shambhala, 1986, pp. 171-181.

4. Krishnamurti, J. *Think on These Things*. New York: Harper & Row, 1964, pp. 45-50.

5. Tart, Charles Ibid, pp. 187-188.

6. Tart, Charles Ibid, pp. 197-198.

Quality Control

1. Sinetar, Marsha *Do What You Love, the Money Will Follow*. Mahwah, NJ: Paulist Press, 1987, pp. 162-192.

2. *California Statistical Abstract*, 1988.

Defining Interior Boundaries

1. Elgin, Duane *Personal Communication*, 1989

Trimmed to Perfection

1. Fuller, Millard and Scott, Diane *Love in the Mortar Joints: the story of Habitat for Humanity*. Piscataway, NJ: New Century Publishers, 1980, pp. 39-53.

2. Maslow, Abraham as described in Tart, *Waking Up*. pg. 173.

3. Anonymous, *A Course in Miracles: Workbook for Students*. Tiburon, CA: Foundation for Inner Peace, 1975, pg.40.

4. Kopp, Sheldon *Raise Your Right Hand Against Fear*. Minneapolis, MN: CompCare Publishers, 1988, pp. 149-150.

5. Ibid. pg. i.

Making Light the Work

1. Capra, Fritjof *The Tao of Physics*. Boulder, CO: Shambhala, 1975, pp. 151-152.
2. Barry, Dave *"How to Make a Board."* New Shelter Magazine (Now called Practical Homeowner), Jan, 1979, pg. 96.

Waste Not, Want Not

1. Fisher, Roger and Brown, Scott *Getting Together: Building a Relationship the gets to Yes*. Boston: Houghton Mifflin, 1988, pg. 32.
2. Ibid. pg. 34.
3. Ibid. pg. 38.

Punching Out

1. Capra, Fritjof *The Turning Point*. in Fields, et al, Chop Wood, Carry Water. Los Angeles, CA: J.P. Tarcher, Inc., 1984, pp. 114-115.
2. Nightingale, Earl *The Strangest Secret*. Cassette Tape, Chicago, IL: Nightingale/Conant, Inc., 1987.
3. Zukav, Gary *The Seat of the Soul*. New York: Simon & Schuster, 1989, pg. 228.

How's the Money

1. Butterworth, Eric *Spiritual Economics*. Unity Village, MO: Unity School of Christianity, 1983, pg. 205.

2. Anonymous *A Course in Miracles: Workbook for Students*. Tiberon, CA: Foundation for Inner Peace, 1975, pg. 8.

Receiving the Fruit

1. Krishnamurti, Jiddu *You Are the World*. New York: Harper & Row, 1972, pg. 158.

Index

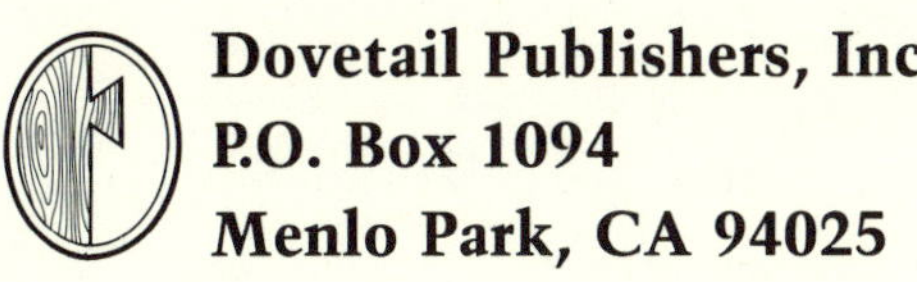

Dovetail Publishers, Inc.
P.O. Box 1094
Menlo Park, CA 94025

Order Form

Name ____

Address ____

City ____

State ____________________ Zip _______________

☐ Check or Money Order Enclosed

Charge my account   Amount _______________

☐ Visa   ☐ Mastercharge   Expiration Date _______________

Card Number _______________________________________

Signature ____

Please send me ____ Copy(s) of *Growing a Housebuilder* at $17.95 plus $2.05 postage and handling. I may examine it for 14 days and if I am not fully satisfied, I may return it undamaged for a full refund.

Note: Books are shipped special 4th class book rate. Please allow four to six weeks for delivery. For RUSH delivery, first class service, please add one dollar extra per book ordered.

A note to the reader:

The author welcomes comments and questions. You may write to him care of the publisher at the address listed above. All correspondence will be responded to. Thank you.

Colophon

This book was written in the early morning hours on a 40 megabyte hard disk residing in the author's Macintosh SE Personal Computer. Copy and structural editing were done by Duane Elgin. Printing was done by Bookcrafters, 613 E. Industrial Drive, Chelsea, Michigan 48118-0370 on 60lb Booktext Natural paper.

Shadow Canyon Graphics, Bryant Drive Office Park, Post Office Box 3699, in Evergreen, Colorado 80439, did the typesetting and book design. The typeface is 11 Point Trump Mediaeval with 3 points of line spacing. It was set using a CRTerminal 300 (front end) and a Linotronic 100 (back end).

Mark Johnson did the ink line drawings.

Spiros Bairaktaris with Electric Art Studios, Mountain View, California, did the cover design and illustration.

The budget for the first printing of 3000 copies was $10,000.